AF690213

GNSS Interference Threats and Countermeasures

For a listing of recent titles in the
Artech House GNSS Technology and Applications Series,
turn to the back of this book.

GNSS Interference Threats and Countermeasures

Fabio Dovis

Editor

ARTECH HOUSE

BOSTON | LONDON
artechhouse.com

Library of Congress Cataloging-in-Publication Data
A catalog record for this book is available from the U.S. Library of Congress

British Library Cataloguing in Publication Data
A catalog record for this book is available from the British Library.

ISBN-13: 978-1-60807-810-3

Cover design by John Gomes

© 2015 Artech House
685 Canton Street
Norwood, MA 02062

All rights reserved. No part of this book may be reproduced or utilized in any form or by any means, electronic or mechanical, including photocopying, recording, or by any information storage and retrieval system, without permission in writing from the publisher.

All terms mentioned in this book that are known to be trademarks or service marks have been appropriately capitalized. Artech House cannot attest to the accuracy of this information. Use of a term in this book should not be regarded as affecting the validity of any trademark or service mark.

10 9 8 7 6 5 4 3 2 1

To my family and friends

Contents

Preface

Writing a book is always a battle between completeness and affordability of the text. This is even more true for a topic such as interference in satellite navigation systems—to provide complete coverage would require detailed descriptions of the theory, of the receiver architectures, and of each method that has been implemented or proposed so far.

The approach followed by the authors in this book has been to find a balance between the two extremes, providing the readers with a fairly complete overview of the different topics, including a good list of references, but at the same time offering insight into the most promising or innovative techniques.

The topic of interference threats is quite hot and new countermeasures are still being proposed, and thus the book focuses on the principles of the methods but avoids providing information that the interested reader can find in the papers included in the references.

Although the first chapter provides some basic principles of satellite navigation receivers, the book is intended for members of the engineering/scientific community who have preexisting knowledge of satellite navigation principles and global navigation satellite systems.

We hope that this book will help engineers and scientists to better understand the interference and spoofing threats, which in turn will help them to design and implement improved robust systems that are able to handle these menaces.

Acknowledgments

When almost 15 years ago I started investigating satellite navigation together with my colleague and friend Paolo Mulassano, I was definitely not expecting that one day in the future, I would have the pleasure of being the editor of a book on satellite navigation topics. For achievement of this milestone I have to thank many people who along the years have given me guidance: colleagues, friends, and students in the NavSAS group. First of all my grateful thanks go to Prof. Letizia Lo Presti, who taught to all of us the passion for research work and the spirit needed to be an effective team. My appreciation is extended in particular to the contributors to this work—Emanuela, Beatrice, Marco, Davide, and Luciano: without your commitment we could have not accomplished the task.

Fabio Dovis
January 2015

1

The Interference Threat

Fabio Dovis

1.1 Introduction to the Book

Reliable positioning and navigation are becoming imperative in more and more applications related to public services, consumer products, and safety-critical situations. Research aimed at finding pervasive and robust positioning methodologies is critical to a growing number of societal areas. Such research also needs to ensure that the navigation is trustworthy and the risks and threats, especially toward satellite navigation, are accounted for. Modern society is highly reliant on global navigation satellite systems (GNSS) and satellite and radio navigation are evolving at an accelerating pace. With the growth of a new European satellite navigation system, Galileo, the development of the Chinese Beidou system, and the modernization of the currently existing systems such as the American GPS or the Russian GLONASS, a wider range of new signals will guarantee better performance, enabling a plethora of new applications. In fact, nowadays, in addition to the obvious usage in positioning and navigation, more and more applications are relying on a robust timing reference from GNSS.

However, although GNSS technology can provide accurate and global positioning, velocity, and time estimations, it is highly vulnerable to a range of threats. GNSS is particularly prone to unintended and malicious radio-frequency interference (RFI) due to the extremely low power level of the signal at the user's receiver after traveling from the satellite transmitter to the receiving antenna on the Earth. Due to the weakness of the GNSS signal that reaches users and a crowded frequency spectrum, GNSS-based services will be always vulnerable to the presence of interfering signals generated by other communication systems. A recent example of these risks was the LightSquared case in the United States, where the GPS receiver operations in the L1 GPS band have been seriously threatened [1].

Furthermore, GNSS threats include intentional attacks with the objective of disrupting the target receiver. Recalling that GNSS bandwidths are protected, the malicious transmission of counterfeit GNSS-like signals, usually known as *spoofing*, may become quite dangerous also for civil use of GNSS. Spoofing and GNSS receiver deception are becoming a threat, as more applications and infrastructures begin to rely on GNSS position and time information. Although the vulnerability of GNSS-based civilian infrastructures is understood, few recognize that severe attacks can be carried out with self-made spoofing devices composed of a software receiver and trivial RF front-ends, as recently demonstrated in the United States by researchers at the University of Austin in Texas [2]. Provision of timing references for communication networks, agriculture, fishery, and road tolling applications are just a few examples of markets that would be deeply affected by spoofing activity designed to elude public authorities or service providers.

Thus, with the growth of civilian GNSS use, unintentional interference, jamming, and spoofing are emerging security challenges in the civil field. There are several applications for which it is essential to detect such types of intentional deception in order to ensure reliable position and time estimations. The provision of such robustness can protect personal safety or infrastructures such as power grids, distribution networks, or communication networks for which GNSS is the provider of timing information. The importance of ensuring a robust receiver with respect to interference and spoofing is crucial for all types of applications where the concept of security is needed. Hence, evaluating the possible impact of potential threats on particular services related to transportation applications (aviation, maritime, railway, road), to emergency applications oriented to the tracking and tracing of sensitive material (e.g., medical or dangerous goods), and to financial/assurance aspects is a priority.

The goal of this book is to provide an overview of the major sources of interference and spoofing for a GNSS receiver, discussing both the methods used to assess their impact on the positioning performance as well as the methods used to protect civilian use of GNSS against unintentional and intentional attacks. This book introduces methods for detection (and possibly mitigation) of intentional and unintentional interference as well as spoofing countermeasures, The techniques investigated in this book have advanced primarily as a result of the increased computational capabilities of GNSS receivers, which allow the implementation of more sophisticated signal processing algorithms with respect to the past. Chipset-based, programmable hardware-based, and fully software-based GNSS receivers are also able to host more complex algorithms for interference mitigation purposes in cases in which it is desirable to mitigate the effect of the interference without discarding the measurements performed. Such algorithms may work at the raw signal sample level, which allows for timely elaboration of warnings and better observability of the phenomenon. The development of innovative algorithms aims at improving the defense mechanisms of several applications and infrastructures with respect to malicious attacks.

1.2 What Is Interference?

It is well known that several phenomena may affect the quality of the pseudorange estimation that is based on the measurement of the propagation time of a signal from a satellite to the user. Any electromagnetic source interacting with the signals is *interfering* with the process of estimating the propagation time. This book focuses on artificial sources of RFI generated either intentionally or unintentionally by some communication system. The following chapters address such sources of artificial interference and the receiver-based techniques used to detect and mitigate their effects. Note, however, that other kinds of interference might be a threat to GNSS positioning performance. They are discussed in the following subsections, but are not be specifically addressed in this book since their detection and mitigation follows specific approaches.

1.2.1 Natural Sources of Interference

When considering the propagation of a signal in the atmosphere, the effect of the ionosphere has to be taken into account due to its impact on the propagation time of the signal. Electron concentration in the ionosphere affects GNSS signals by introducing delays in their propagation. Such errors can be corrected in part by making use of models of the background ionosphere when performing single-frequency measurements, or corrected entirely in the case of dual-frequency measurements. However, in some cases electron density irregularities may appear that can further disrupt the propagation of the wave by introducing fluctuations in amplitude and phase; such phenomena are usually called *scintillations* [3]. How often GNSS signals are affected by scintillations depends on solar and geomagnetic activity, geographic location, season, local time, and signal frequency. Scintillation can be considered a sort of natural interference interacting with the GNSS signal that causes the signals to fade and induces a frequency shift in the signal carrier that in some cases can strongly affect the GNSS receiver. During strong ionospheric events, amplitude fades and frequency variations can be very challenging for a receiver and may cause frequent cycle slips and losses of lock of the satellite signals [4].

1.2.2 Multipath

Multipath occurs whenever the user device receives reflected signals in addition to the direct line-of-sight signal. These replicas of the signals are generated from the ground, buildings, or trees in terrestrial navigation, whereas signal reflections from the host-vehicle body are more common in airborne and marine applications. Multipath can be *specular* when generated from smooth surfaces or *diffuse* when arising from diffuse scatterers and sources of diffraction.

To a certain extent, multipath can then be considered a self-interference, where the interfering signal is a replica of the signal itself.

1.2.3 Intersystem and Intrasystem Interference

The signal impinging the GNSS receiver antenna at a given frequency is the combination of the signals broadcast by all the satellites in view. GNSS RF compatibility addresses the issue of intrasystem (from the same system) and intersystem (from other systems) interference. Signals belonging to the same satellite constellation are designed to be theoretically orthogonal (exploiting code or frequency diversity), and thus they can be separated by the receiver processing. However, such orthogonality is not perfect and a residual power is always generating intrasystem interference.

Intersystem interference is due to the fact that several GNSS systems share the same carriers, and again, some power from the signals of another system can disrupt the signal of interest. Several methodologies, such as the effective carrier power to noise density theory introduced in [5], are used during the design phase of the systems to ensure that a maximum acceptable level of intersystem interference is respected (see, e.g., [6, 7]). Intra- and intersystem interference is then a topic that needs to be addressed during the design phase, and it is beyond the regular users' capabilities to deal with it.

Due to the growing number of operational satellites in the new GNSS constellations, the number of satellites that are in view to a user receiver at the same time is growing as well. This implies increased intersystem interference. However, from a user's standpoint, it can only be reduced by means of directional antennas that can spatially filter the signals coming from satellites that are not of interest.

1.2.4 Artificial Interference: Unintentional and Intentional Interference

The intrinsic power weakness of GNSS signals affects the performance of any type of receiver, since all the communication systems transmitting at carrier frequencies close to the band of interest are potential sources of interference for a GNSS receiver, and even small leakages out of their allocated bandwidth can be threatening to GNSS signals. Even though unintentional RFI events are generally unpredictable, their presence has been experienced in the past and the increasing number of wireless communication infrastructures is increasing the probability that some power spillover from signal frequencies located near the GNSS bands could affect the performance of GNSS receivers in a certain region. The presence of interfering power can be due to several reasons, but the main effects are caused by harmonics or spurious components generated by intermodulation products in the communication transmitter.

Jamming refers to intentional transmission of RF energy to hinder a navigation service by masking GNSS signals with noise. The malicious objective of jammers is to cause the receiver to lose tracking and to impede signal reacquisition. Although jamming is a well-known threat in the military applications, it represents a growing threat for many GNSS-based applications. Systems involving safety and liability-critical operations (e.g., safe navigation in ports, systems for smart parking and tolling, GNSS-based synchronization of power networks) could potentially be heavily impaired by jamming attacks. The level of threat associated with jamming cannot be disregarded, considering that portable jammers are available online and can be purchased at a very low cost. Although the use of jammers is not legal, the interest of individuals willing to break the law may result in fraudulent actions toward GNSS-enabled systems. Several studies have addressed the characterization of commercial jammers and their effect on GNSS receivers, demonstrating that they can affect GPS receivers' functionality even if located up to 9 km away (see, e.g., [8, 9]). The intentional transmission of a GNSS-like signal is referred to as *spoofing*, to distinguish the transmission of specific signals aimed at disrupting the operations of the receivers from the generic introduction of in-band powerful disturbances. More details about spoofing techniques are provided in Chapter 3.

1.3 Does Radio-Frequency Interference Exist?

Coffed [10] writes that "Although GPS jamming incidents are relatively rare they can occur; and, when they do, their impact can be severe." In fact, nowadays topics related to security aspects are very hot in the GNSS community and very recent publications, even contemporary to the time of this writing [11], can be found. On February 13, 2014, the *Financial Times* published an interview with one of the GPS founders, Professor Bradford Parkinson [12], on the security of systems relying on GPS. Professor Parkinson clearly recalled the challenge of making GNSS-based systems more robust. For example, cell phone towers are often timed with GPS and if they lose their timing reference, the network loses synchronization with a consequent risk of loss of service. Professor Parkinson also referred to these concepts during his keynote speech titled "Assured PNT—Assured World Economic Benefits" at the European Navigation Conference ENC-GNSS 2014, where he presented his proposal in response to the GNSS vulnerabilities [13, 14].

The concerns of Prof. Parkinson are shared by many GNSS experts. It is in fact clear that interference is one of the main limitations to the development of GNSS-based applications and services. The threat is relevant when the interference is unpredictable, because in other cases the receiver can implement an ad hoc solution for specific interfering sources, as is the case for the aeronautical bandwidths that are shared with other radio-aiding communication systems. In recent times, several unexpected interference events have been reported; for the sake of presenting an example, some of them are briefly described in the following subsection.

1.3.1 Examples of Real Cases of RF Interference

Some of the literature reports about GPS failures that occurred during trials and/or experiments in controlled interference scenarios. Many other works report cases of GPS failures in real situations. Some examples of both testing results and interference incidents are as follows:

- In January 2007, GPS services were significantly disrupted throughout San Diego, California [11]. Naval Medical Center emergency pagers stopped working, the harbor traffic-management system used for guiding boats failed, airport traffic control had to use backup systems and processes to maintain air traffic flow, cell phones users found they had no signal, and bank customers trying to withdraw cash from automated teller machines (ATMs) were refused. It took 3 days to find an explanation for this mysterious event: Two Navy ships in San Diego Harbor had been conducting a training exercise when technicians jammed radio signals. Unwittingly, they also blocked GPS signals across a broad swath of the city [11].

- A famous incident, well known in the GNSS community, occurred at Newark Airport, New Jersey, in 2010 when one of the local-area augmentation system (LAAS) ground facility (LGA) receivers was occasionally jammed by personal privacy devices (PPDs) installed onboard vehicles passing along a nearby motorway. In that case, some of the truck drivers were illegally using a jammer to inherit the GNSS receiver and hide their trucks' positions from the truck fleet manager. The use of GNSS jammers is currently growing in the road domain and starting to be tackled. This event is also meaningful due to the effort required to determine that emissions from mobile PPDs were responsible for the interference at Newark Airport [11, 15].
  Eventually, in August 2013, the Federal Communications Commission (FCC) fined a man nearly $32,000 (Readington, New Jersey) after concluding he interfered with Newark Liberty International Airport's satellite-based tracking system by using an illegal GPS jamming device in his pickup truck to hide from his employer. The signals emanating from the vehicle blocked the reception of GPS signals used by the air traffic control system.
- In January 2011, the U.S. FCC waived restrictions against terrestrial transmitters in the 1525–1559-MHz band allocated for space-to-Earth satellite communications. The agency issued an order that allowed LightSquared Subsidiary LLC to proceed with its plan to deploy a network of base stations, under the condition that the company form a working group to look into the GPS interference issue [1, 17]. The report of the Technical Working Group (TWG) was submitted to the FCC on June 30, 2011, demonstrating widespread adverse effects by LightSquared transmissions on all categories of receivers tested [18]. Wideband receivers, in particular, seem to be adversely affected by the adjacent LightSquared interference; this fact has worried the military community and the civil high-precision applications stakeholders.

- An interesting description of a trial conducted in 2008 on GPS jamming in the maritime sector can be found in [10]. It perfectly highlights how a GPS denial might strongly affect other onboard equipment. The experiment was conducted by the General Lighthouse Authorities of the United Kingdom and Ireland (GLAs), in collaboration with the U.K. government's Defence Science and Technology Laboratory (DSTL) at Flamborough Head on the east coast of the United Kingdom. A low-to-medium power jammer, controlled remotely by two very-high-frequency (VHF) transceivers, transmitted a known pseudorandom noise code over the civilian L1 frequency, which provided a jamming signal over the whole 2-MHz bandwidth of L1, and a trial vessel made several runs between two waypoints positioned outside the jamming area. Authors of [19] outline all of the direct and indirect effects that the GPS jamming unit had on both the onboard equipment and the reference station. Among the onboard equipment, GPS and eLoran receivers, automatic identification systems (AIS), digital selective calling (DSC) systems, and the vessel's electronic chart display & information system (ECDIS) manifested some malfunctioning. Onshore, the differential GPS (DGPS) reference station and the synchronized lights (conventional aid-to-navigation systems) were affected by the presence of the jammer.
- Two interference events due to spurious emission of TV transmitters were detected in 2006. In one case [20, 21], the disturbance, likely due to digital video broadcasting television (DVB-T) transmitters, was the cause of performance degradation in the acquisition stage of a GPS receiver operating in the area, with a consequent loss of the GPS signal tracking. In the latter case, ultrahigh-frequency (UHF) harmonics have been detected in Sydney, Australia, around TV antennas. The undesired signal in the L1 band corrupted the correct performance of the receiver chain, leading to significant variations in the AGC/ADC block and in the final user positioning [22].

This list of events is, of course, not exhaustive and further examples of real cases of interference events can be found, for example, in [23].

1.4 Review of Digital GNSS Receivers

A full description of the GNSS receiver architecture is beyond the scope of this book. However, we discuss here the main aspects related to the receiver and to the signal model because such a discussion will prove useful in the following chapters.

In Figure 1.1 a simplified scheme of the first operational stages of a GNSS receiver is illustrated. The received signal $y_{\mathrm{RF}}(t)$ is composed of the sum of all received waveforms broadcast by the N_S satellites in view at the time of measurement, noise, and other disturbing signals and can be written as

$$y_{\mathrm{RF}}(t) = \sum_{l=0}^{N_S-1} s_{\mathrm{RF},l}(t) + i(t) + n(t) \tag{1.1}$$

where $s_{\mathrm{RF},l}(t)$ is the useful GNSS signal received by the lth satellite in line of sight, $i(t)$ is the additive interfering signal transmitted over a carrier frequency f_{int} and characterized by a two-sided bandwidth B_{int}, and $n(t)$ is the additive white Gaussian noise.

The front-end block is in charge of demodulating the composite received signal to an intermediate frequency (IF) and passing it through a filter with bandwidth B_{IF} to remove the image frequencies. At the output of the ADC/AGC block of Figure 1.1, composed of the analog-to-digital converter (ADC) driven by the automatic gain control (AGC), the continuous signal is digitized in $y_{\mathrm{IF}}(nT_S)$, where $T_s = 1/f_s$ is the time sampling interval, and n is the discrete-time index. Thus, the composite received signal at the ADC/AGC output can be written as

$$y_{\mathrm{IF}}[n] = y_{\mathrm{IF}}(nT_s) = Q_k^u\left[\sum_{l=0}^{L-1} s_{\mathrm{IF},l}(nT_s) + i_{\mathrm{IF}}(nT_s) + \eta(nT_s)\right] \tag{1.2}$$

where $i_{\mathrm{IF}}(t)$ is the demodulated version of the interfering signal (filtered if $B_{\mathrm{int}} > B_{\mathrm{IF}}$) and $\eta(t)$ is the filtered Gaussian noise, the function Q_k^u denotes the quantization over k bits, and T_s is the sampling interval. Expanding the term

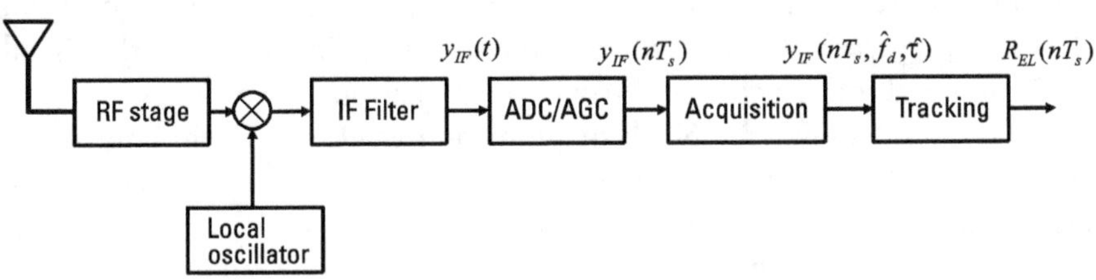

Figure 1.1 Functional blocks of GNSS receiver.

$S_{\mathrm{IF},l}(nT_s)$, the expression for the single digitized GNSS signal affected by noise and interference components becomes (neglecting for the sake of simplicity the subscript l)

$$y_{\mathrm{IF}}[n] = Q_k^u\left[\sqrt{2C}d\left[n-n_0\right]c\left[n-n_0\right]\cdot\cos\left(2\pi F_{D,0}n+\varphi_0\right)+i_{\mathrm{IF}}[n]+\eta[n]\right] \quad (1.3)$$

where C is the received GNSS signal power from one satellite in view, $d[n]$ and $c[n]$ are, respectively, the navigation data message signal and the pseudorandom noise sequence, $F_{D,0} = (f_{\mathrm{IF}} + f_0)T_s$ is the Doppler-affected frequency, $n_0 = (\tau_0/T_s)$ is the digital code delay, ϕ_0 is the instantaneous carrier phase, and $i[n]$ and $\eta\,[n]$ are the digitized interference and the digital Gaussian noise component, respectively. Given B_{IF}, the front-end bandwidth, it can be shown that after sampling the signal at the Nyquist frequency $f_s = 2B_{\mathrm{IF}}$, the noise variance becomes

$$\sigma_{\mathrm{IF}}^2 = E\left\{\eta^2[n]\right\} = \frac{N_0 f_s}{2} = N_0 B_{\mathrm{IF}} \quad (1.4)$$

where $N_0/2$ is the power spectral density (PSD) of the noise.

In the acquisition block, Doppler frequency $\hat{f}_d$ and code phase $\hat{\tau}$ estimations are provided by correlations among the in-phase and quadrature components of the incoming signal and a GNSS code local replica. More details about the acquisition procedure are available, for example, in [24, 25] and are not addressed further in this chapter. The effect of the different types of interference on the acquisition stages is investigated in Chapter 2.

The signal tracking follows the signal acquisition. Over each channel of the receiver, a delay lock loop (DLL) is used to synchronize the received spreading code and a local replica, while a phase lock loop (PLL) is generally employed to track the phase of the incoming carrier. The signal tracking relies on the properties of the signal correlation and is fundamental to demodulate the navigation message and estimate the range between the user and the satellites. Conventional receiver architectures generally include a frequency lock loop (FLL) to refine the rough estimate performed by the signal acquisition. The FLL eases the PLL lock, reducing the transient time between the signal acquisition and the steady-state carrier/code tracking.

Figure 1.2 shows the block diagram of a tracking system commonly used in digital GNSS receivers for a single channel, but the same architecture is repeated over all channels to track different satellites (or different channels from the same satellite in case of composite signals as foreseen for the Galileo system).

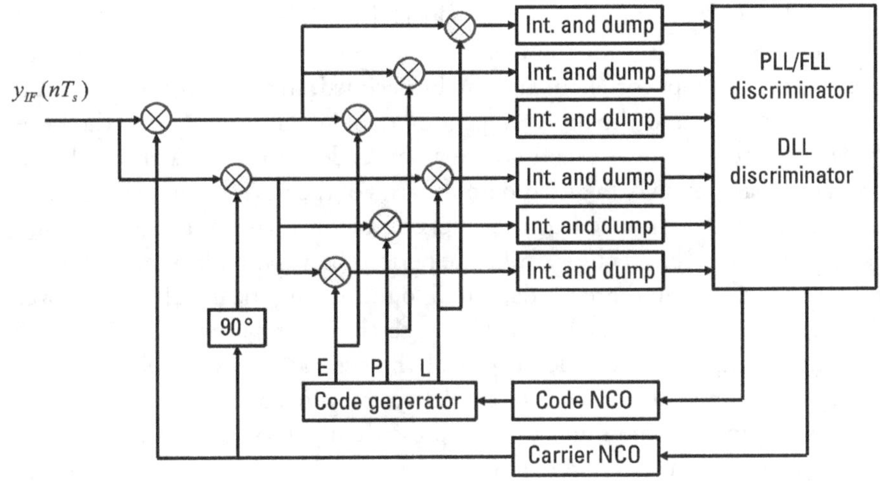

Figure 1.2 Block diagram of a code and carrier tracking loop for GNSS receivers.

The tracking loop relies on correlation operations between the received signal and local replicas of carrier and code, initialized by the Doppler frequency $\hat{f}_d$ and the code phase $\hat{\tau}$ estimated in the acquisition phase.

The values of correlation are then used to produce feedback control signals on the basis of proper discrimination functions; one for the PLL and one for the DLL. Such control signals are filtered and used to steer the code and carrier generators that prepare the local replicas for the next loop iteration. The process continues and the system follows the input signal variations over time. Note that the described synchronization process corresponds to finding the best estimate of the local carrier frequency/phase and local code delay that maximize the correlation between the incoming and the local replicas.

Noncoherent tracking systems, like that shown in Figure 1.2, use two branches, one in phase (I) and the other in quadrature (Q). Generally speaking, noncoherent tracking loops are more robust and do not require the estimate of the carrier phase (i.e., they do not necessarily need a PLL; an effective system can be designed combining an FLL and a DLL). For example, right after the signal acquisition, when the tracking phase starts, the system has not yet recovered the phase of the incoming carrier and part of the power goes on the quadrature branch. Different from coherent tracking loops (that use only the I branch), in this case, using both the branches, the discriminators are still able to produce feedback signals. If a PLL is used, after an initial transient time, the incoming carrier is synchronized with the local one and the received signal is completely converted on the I branch.

When both the DLL and PLL are locked, the incoming signal is despread and converted to baseband. The navigation data bits appear at the output of the in-phase prompt correlator and can be decoded. In addition, with the DLL locked, the local and the incoming codes are aligned. Referring to the local code, the receiver knows exactly when a new code period starts and is able to recognize navigation data bits and boundaries of the navigation message. The receiver stays synchronized to the tracked satellites, continuously counting the number of received chips, full code periods, navigation bits, and message frames. These counters are fundamental to measuring the misalignment over different channels and tracking different satellites, and are used to compute the pseudoranges. Once at least four of the pseudoranges are obtained the position is estimated by means of a trilateration procedure.

In Chapter 2, the impact of the presence of $i(t)$ on the different stages of the receiver is analyzed, showing the effect on the acquisition probabilities and on the tracking jitter.

1.5 Organization of the Book

The book is divided into two parts. Chapters 1, 2, and 3 provide an overview and classification of interference and spoofing sources. The different sources are discussed in terms of their features (frequency, modulation, and so on) and their proper model with respect to the GNSS signals. Chapter 4 introduces some common techniques for the analytical assessment of the interference effects, and can be used as a reference for the prediction of the performance of a GNSS receiver in an interfered environment.

The second part of the book is then devoted to describing the techniques for the detection and mitigation of interference and spoofing attacks. Chapter 5 presents an overview of the common interference detection techniques tailored to the different families of interference. Mitigation of interference is addressed by Chapters 6 and 7, presenting classical mitigation techniques and advanced signal processing techniques, respectively. Chapter 8 discusses the best strategies for providing antispoofing features to GNSS civil signals.

References

[1] http://www.gps.gov/spectrum/lightsquared/.

[2] Humphreys T., et al., "Assessing the Spoofing Threat: Development of a Portable GPS Civilian Spoofer," *in Proc. of the 21st International Technical Meeting of the Satellite*

Division of the Institute of Navigation (ION GNSS 2008), Savannah, GA, September 2008, pp. 2314–2325.

[3] Yeh, K. C., and C.-H. Liu, "Radio Wave Scintillations in the Ionosphere," *Proc. IEEE*, Vol. 70, No. 4, 1982, pp. 324–360.

[4] Doherty, P. H., et al., "Ionospheric Scintillation Effects in the Equatorial and Auroral Regions," *Proc. 13th Int. Technical Meeting of the Satellite Division of the Institute of Navigation (ION GPS 2000)*, Salt Lake City, UT, pp. 662–671.

[5] Betz, J. W., "Effect of Narrowband Interference on GPS Code Tracking Accuracy," *Proc. 2000 National Technical Meeting of the Institute of Navigation*, Anaheim, CA, January 2000, pp. 16–27.

[6] Titus, L. B. M., et al., "Intersystem and Intrasystem Interference Analysis Methodology," in *Proc. ION GPS/GNSS 2003*, Portland, OR, September 2003.

[7] Liu, W., et al., "GNSS RF Compatibility Assessment: Interference Among GPS, Galileo, and Compass," *GPS World*, December 2010.

[8] Mitch, R. H., et al., "Civilian GPS Jammer Signal Tracking and Geolocation," *Proc 25th Int. Technical Meeting of The Satellite Division of the Institute of Navigation (ION GNSS 2012)*, Nashville, TN, September 2012, pp. 2901–2920.

[9] Borio, D., C. O'Driscoll, and J. Fortuny, "Jammer Impact on Galileo and GPS Receivers," *Proc. 2013 Int. Conf. on Localization and GNSS (ICL-GNSS)*, June 25–27, 2013, pp. 1, 6. doi:10.1109/ICL-GNSS.2013.6577265

[10] Grant, A., and P. Williams, "GNSS Solutions: GPS Jamming and Linear Carrier Phase Combination," *Inside GNSS*, Vol. 4, No. 1, January/February 2009.

[11] Coffed, J., "The Threat of GPS Jamming. The risk to an Information Utility"; available at http://www.exelisinc.com/solutions/signalsentry/Documents/ThreatOfGPSJamming_February2014.pdf.

[12] Jones S., and Hoyos C.,"GPS Pioneer Warns on Network's Security," *Financial Times*, http:// http://www.ft.com/cms/s/0/fadf1714-940d-11e3-bf0c-00144feab7de.html#axzz3J2VEueWr.

[13] Gutierrez, P., "At ENC 2014: A GNSS Wake Up Call for Europe," *Inside GNSS News*, April 16, 2014; available at http://www.insidegnss.com/node/3985.

[14] Jewell, D., "Protect, Toughen, Augment: Words to the Wise from GPS Founder," *GPS World*, April 15, 2014; available at http://gpsworld.com/protect-toughen-augment-words-to-the-wise-from-gps-founder.

[15] Grabowsky, J. C., "Personal Privacy Jammers. Locating Jersey PPDs Jamming GBAS Safety-of-Life Signals," *GPS World*, Vol. 23, No. 4, April 2012.

[16] Pullen, S., and G. X. Gao, "GNSS Jamming in the Name of Privacy," *Inside GNSS*, Vol. 7, No. 2, March/April 2012.

[17] "LightSquared Fails FCC GPS Interference Tests," 360 Degrees Column, *Inside GNSS*, Vol. 6, No. 4, July/August 2011, pp. 12–15.

[18] Boulton, R., et al., "GPS Interference Testing—Lab, Live, and LightSquared," *Inside GNSS*, Vol. 6, No. 4, July/August 2011, pp. 32–45.

[19] Grant, A., et al., "GPS Jamming and the Impact on Maritime Navigation," *Journal of Navigation*, Vol. 62, No. 2, April 2009, pp 173–187.

[20] Motella, B., M. Pini, and F. Dovis, "Investigation on the Effect of Strong Out-of-Band Signals on Global Navigation Satellite Systems Receivers," *GPS Solutions*, Vol. 12, No. 2, March 2008, pp. 77–86.

[21] De Bakker, P., et al., "Effect of Radio Frequency Interference on GNSS Receiver Output," *Proc. 3rd ESA Workshop on Satellite Navigation User Equipment Technologies (NAVITEC 2006)*, ESA/ESTEC, Noordwijk, The Netherlands, December 2006.

[22] Balaei, A. T., B. Motella, and A. G. Dempster, "GPS Interference Detected in Sydney-Australia," *Proc. 2007 Int. Global Navigation Satellite System (IGNSS 2007) Conf.*, Sydney, Australia, December 2007.

[23] Motella, B., et al., "Assessing GPS Robustness in Presence of Communication Signals," *Communications Workshops 2009*, June 14–18, 2009, pp. 1, 5. doi:10.1109/ICCW.2009.5207985

[24] Kaplan, E., and C. Hegarty, *Understanding GPS Principles and Applications*, 2nd ed., Norwood, MA: Artech House, 2005.

[25] Misra, P., and P. Enge, *Global Positioning System: Signals, Measurements, and Performance*, Lincoln, MA: Ganga-Jamuna Press, 2006.

2

Classification of Interfering Sources and Analysis of the Effects on GNSS Receivers

Fabio Dovis, Luciano Musumeci, Beatrice Motella, and Emanuela Falletti

2.1 Introduction

A global navigation satellite system (GNSS) receiver is vulnerable to several kinds of radio-frequency interference (RFI) due to the fact that it has to extract pseudorange information by processing the signal in space (SIS), which is received at a very low signal power.

The nominal received power is on the order of magnitude of −160 dBW for all GNSSs, without taking into account extra attenuations that may be due to the local environment. Despite the weakness of the signals, the spread-spectrum nature of the SIS allows navigation receivers to recover timing information and to estimate the pseudoranges necessary to compute the user's position by exploiting the gain obtained at the output of the correlation block. Even if the correlation process is theoretically able to mitigate the presence of nuisances in the bandwidth of interest, a real limitation can be the finite dynamic range of the receiver front-end. The presence of undesired RFI and other channel impairments can result in degraded navigation accuracy or, in severe cases, in a complete loss of signal tracking.

This chapter introduces a general classification of the interfering sources, presenting an overview of the main terrestrial systems that are potential sources of RFI for the GNSS signals. The second part of the chapter discusses their effect on the different stages of the GNSS receiver.

2.2 Classification of Interfering Source

The classification of the main disturbances for GNSS receivers takes into account heterogeneous aspects. The emission types can be defined as being intentional (jamming) or unintentional as described in Chapter 1. The first are common for military scenarios even if jamming of civil applications starts to be common due to the availability of jamming devices on the market.

Furthermore, a large number of communication systems present in our daily lives emit power that could interfere with the GNSS L-band, due to out-of-band emissions by these electronic systems.

We turn now to a discussion of the classification of interfering sources, based on their spectral and time features.

2.2.1 Interference Spectral Features

A general classification of the interfering signals is based on their spectral characteristics such as carrier frequency f_{int} and bandwidth B_{int}, with respect to the GNSS signal carrier f_{GNSS} and occupied bandwidth B_{GNSS}

- *Out-of-band interference* refers to interfering signals whose carrier frequency is located near to the targeted GNSS frequency band ($f_{\text{int}} < f_{\text{GNSS}} - B_{\text{GNSS}}/2$ or $f_{\text{int}} > f_{\text{GNSS}} + B_{\text{GNSS}}/2$.
- *In-band interference* refers to interfering signals with carrier frequency within the GNSS frequency band ($f_{\text{GNSS}} - B_{\text{GNSS}}/2 < f_{\text{int}} < f_{\text{GNSS}} + B_{\text{GNSS}}/2$).

Moreover, interference can be further classified according to its characteristics in the frequency domain as follows:

- *Narrowband interference (NBI):* The spectral occupation is smaller with respect to the GNSS signal bandwidth ($B_{\text{int}} \ll B_{\text{GNSS}}$).

- *Wideband interference (WBI):* The spectral occupation is comparable with respect to the GNSS signal bandwidth ($B_{int} \approx B_{GNSS}$).
- *Continuous-wave interference (CWI):* Represents the ultimate limit in NBI and appears as a single tone in the frequency domain ($B_{int} \rightarrow 0$).

Furthermore, in general, interference might have frequency-varying characteristics, for example, the chirp signals characterized by a linear variation in time of the instantaneous frequency, thus appearing as WBI. This kind of interfering signal is typically generated by the jammers. Such devices are capable of transmitting strong power chirp signals sweeping several megahertz in a few microseconds, thus obscuring the correct signal reception of each GNSS channel. Due to their availability on the web, this type of intentional interfering signal is gaining more and more attention in civilian applications.

CWI could have a severe impact on a GNSS receiver, either on the acquisition or on the tracking process, because the interference power is dispersed on the whole search space by the correlation with the local code, compromising the acquisition accuracy and affecting the other functional blocks. The impact of CWI and NBI strongly depends on the value of the central frequency of the interference within the frequency band. This is due to the almost periodic nature of GNSS signals. In fact, the spectrum of a GNSS signal has components spaced at multiples of the inverse of the code period (e.g., 1 kHz for GPS C/A code) with different power allocated to each component depending on the shape of the code spectrum. The impact of CWI is larger in cases where the CWI is matched with such components [1–3].

2.2.2 Pulsed Interference

Pulsed interfering signals are characterized by an on–off status of short duration (order of microseconds), which alternate in the time domain. This type of interference signal is typical of aviation scenarios, where several aeronautical radio navigation Services (ARNS) broadcast strong pulsed signals in a bandwidth that is shared with some of the satellite navigation systems.

The parameters used to describe pulsed interference are:

- *Pulse width (PW):* duration of one pulse;
- *Pulse repetition frequency (PRF):* number of pulses per second;
- *Duty cycle (DC = PRF*PW):* the percentage of time that is occupied by the pulses.

Pulsed interference with low DC has a small impact on receiver performance compared to continuous interference with the same power and center frequency.

2.3 Potential Interference Sources

The potential interference can share the GNSS frequencies (in-band RFI) or be far from the GNSS carrier (out-of-band RFI). There are almost no in-band authorized emissions in the GNSS bandwidths; however, interference comes mainly from the spurious emissions of out-of-band systems, which generate harmonics that collide with the GNSS bandwidths.

2.3.1 Out-of-Band Signals

In the following sections, some of the main potential out-of-band interference sources are analyzed.

Analog TV Channels

TV emissions are veritable sources of interference for a GNSS receiver. They can manifest as both wideband and narrowband interference: The video carriers are considered to be medium/wideband signals, whereas the sound carriers are considered to be CWI. In the broadcast TV signal, VHF and UHF bands are used. The harmonics of such bands generated by TV ground station transmitters can generate potentially dangerous interference for GNSS receivers as depicted in Figure 2.1.

As an example, in [4] a case of interference from a TV signal is reported. In this case, the interference signal affects the active antenna LNA causing harmonic distortion in the same LNA that results in an average of 5-dB loss in C/N_0. In [3] six TV channels, French and American equivalents, with their harmonics are analyzed in frequency and power terms.

DVB-T Signals

The DVB standard has been defined (since 1993) within an initiative involving more than 300 European and extra-European members. The DVB project harmonized the strategies for introducing digital television and the new multimedia interactive services on transmission networks. It also defined the technical specifications. The project defined the system specifications for standard Digital Video Broadcasting–Satellite (DVB-S), developed for the direct diffusion of TV multiprogramming from satellites and for standard Digital

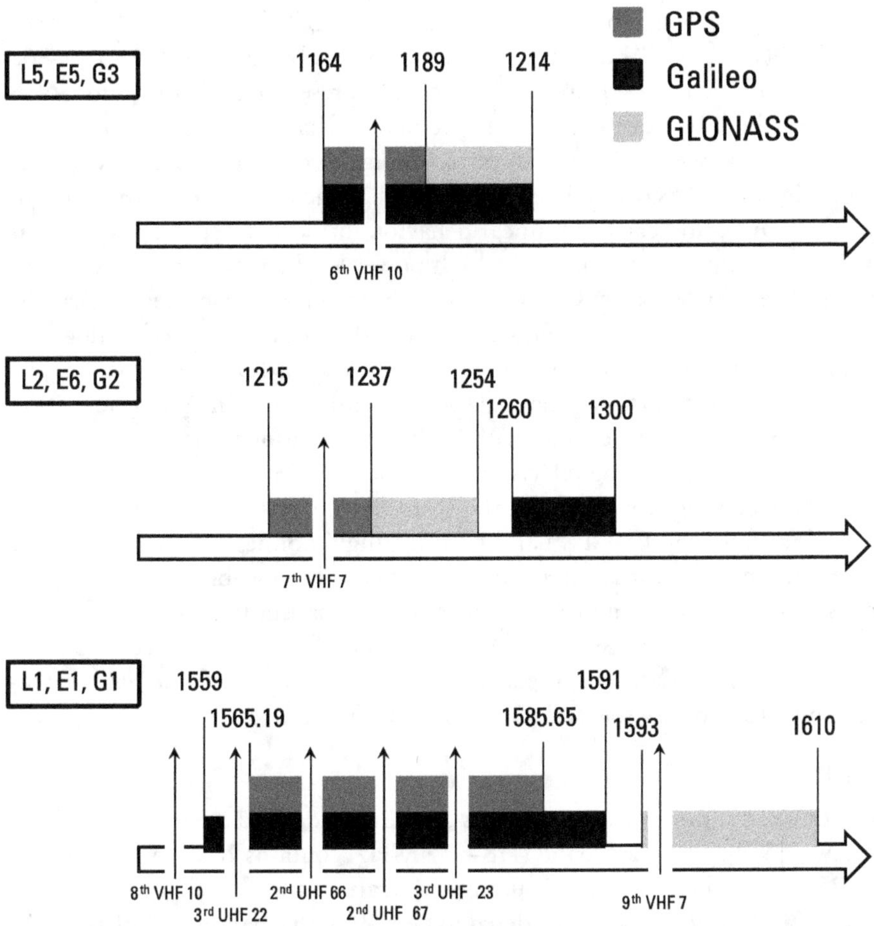

Figure 2.1 Potential TV channel harmonic interference.

Video Broadcasting–Cable (DVB-C) for the distribution of television signals through core networks. The DVB family also comprehends the standard for Digital Video Broadcasting–Terrestrial (DVB-T), for the provision of wireless digital terrestrial television. The DVB-T standard is based on the Moving Pictures Experts Group-2 (MPEG-2) standard for audio/video signal source coding and it adopts a multi-carrier modulation COFDM to distribute the total data stream among a large number of carrier frequencies equally spaced and modulated using QPSK, 16-QAM, 64-QAM, nonuniform 16-QAM, or nonuniform 64-QAM [5].

In the European broadcasting area, the DVB-T frequency bands are the VHF III (174–230 MHz), UHF IV (470–862 MHz), and UHF V (582–862 MHz) bands. These frequency values do not represent a direct threat to GNSS receivers, but they can cause some problems if harmonics due to potential distortions caused by the malfunction of some electronic devices, like power amplifiers, are considered. Even one single damaged amplifier in the amplification chain could cause nonlinear behavior, introducing spurious emissions at the RF output that, due to the high power level emitted, could represent a real threat for a nearby GNSS receiver. Furthermore, considering that the frequency involved in the DVB-T signal is the same of that of analog TV, the probability of having some disturbances caused by DVB-T signals can be considered similar to that of having spurious emissions from analog television systems. In [6–8] some examples of significant variations in the quality of the GPS signal due to analog television transmitters are reported.

Considering, for example, the third harmonics[1] of UHF V carrier, it would fall into the L1 GPS band representing a nonnegligible threat to the receiver. Therefore, it is important to evaluate the possibility of distortions caused by nonlinear amplifiers or linear ones in saturation.

A detailed analysis of OFDM DVB-T potential interference in Europe is reported in [7] where the impact of RFI on the GNSS useful signal is evaluated by means of the spectral separation coefficient.

VHFCOM

Other VHF communication systems can be considered dangerous to a GNSS receiver [3, 9]. The VHF band (118–137 MHz) contains 760 channels spaced by 25 kHz, and it is commonly used by air traffic control (ATC) communications. The harmonics are considered to be NBI with a bandwidth of about 25 kHz. The VHF channels, centered at 121.150, 121.175, and 121.200 MHz, have the 13th harmonic within the GPS bandwidth, whereas the channels centered at 131.200, 131.250, and 131.300 MHz are dangerous for their 12th harmonic. In Figure 2.2 VHF communication (VHFCOM) potential harmonics are depicted.

FM Harmonics

Also small frequency bands inside the FM band (87.5–108 MHz) have harmonics that fall in the GNSS bands. The channels at 104.9 and 105.1 MHz have their 15th harmonics near the GPS and Galileo bandwidths as depicted

[1] The harmonic order is considered with respect to the signal fundamental frequency (f_0) adopting the definition used in [8].

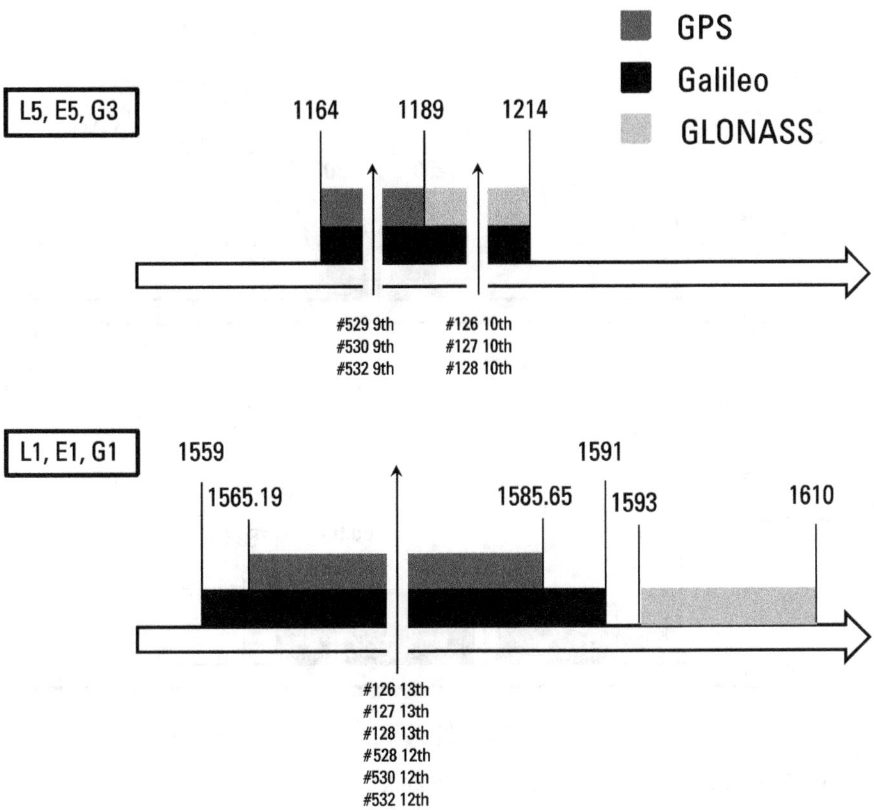

Figure 2.2 Potential VHFCOM channel harmonic interference.

in Figure 2.3. The channels are spaced at 150 kHz, while the maximum transmitted power is 50 dBW. The harmonics generated by FM sources are considered as WBI with respect to GNSS signals allocated in the L1/E1 bands.

Personal Electronics Devices

Personal electronic devices (PEDs) in proximity to a GNSS receiver can cause the disruption of GNSS signal reception. PEDs include cell phones, pagers, two-way radios, remote control toys, laptops, and many others. A larger number of PEDs are expected to include, in the future, ultra-wideband (UWB) transmission that allows the development of high-bit-rate personal devices.

SATCOM

Satellite communications (SATCOM) operate in the frequency bands of 1626–1660.5 MHz with channels spaced at 0.75 MHz and bandwidth of 20 kHz.

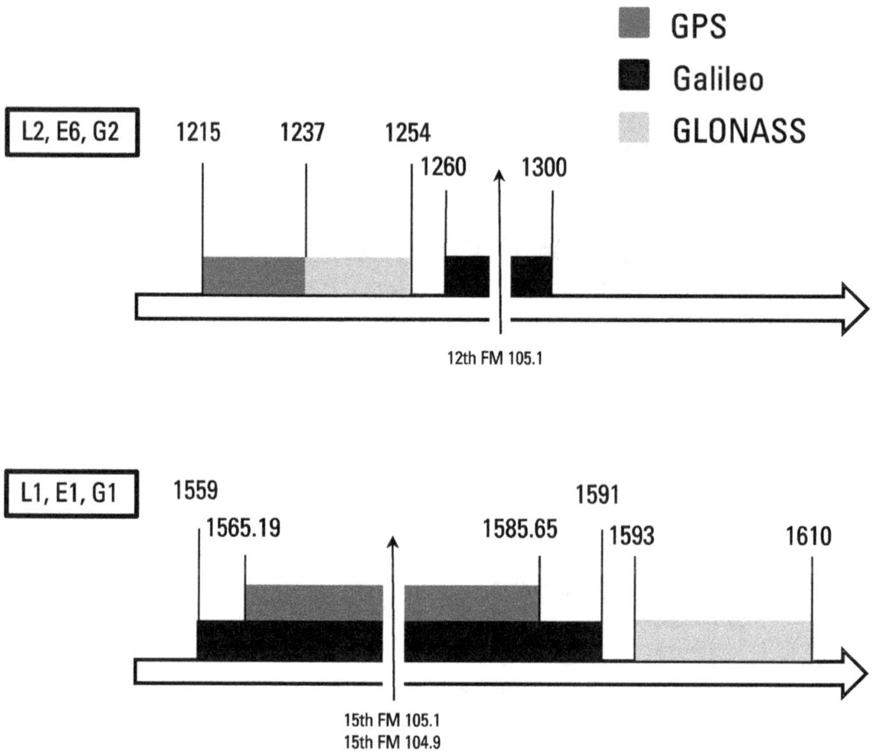

Figure 2.3 Potential FM harmonic interference.

Multi-carrier transmission in a SATCOM service produces intermodulation products that can fall in the GNSS band. A possible example is reported in [3].

VOR and ILS Harmonics

The VHF omnidirectional range (VOR) is a type of radio navigation system for aircraft that provides information about radial position with respect to ground station. The instrument landing system (ILS) consists of two radio transmitters providing lateral and vertical guidance to aircraft for approaching landing. VOR/ILS emitters are usually positioned at the beginning, end, and sides of airport runways. These approaching landing systems operate in the 108–117.95-MHz band including 200 channels frequency spaced at 50 kHz. In detail, the VOR uses 12 channels in the 112.24–112.816-MHz band while the ILS localizer transponder only uses one frequency on 40 channels in the 108.10- to 111.95-MHz band. Their harmonics, the 14th from VOR and 2nd from ILS corresponding to 111.9 and 111.95 MHz, enter on the L1/E1 band. They are considered CWI signals.

Mobile Satellite Service (MSS)

The mobile satellite service (MSS) system can generate two distinct interference threats to a GNSS receiver. The MSS mobile earth stations use the 1610–1660.5-MHz band, potentially introducing wideband power in the GNSS band.

Mobile Phone Interference

In general, no direct consequences from mobile phones on a GNSS receiver have been reported in the literature so far. Some information is available for aircraft navigation equipment, where a GPS receiver is commonly used. In [10], an investigation of spurious emissions from six wireless phone technologies is described, analyzing the effects on aircraft systems, among which is a GPS. The tests are carried out from semianechoic and reverberation chambers using wireless phone technologies with different transmission frequencies and different receiving antennas. The study is conducted evaluating the total radiated power from each cellular versus the frequencies of the system considered. In the analysis, the receiver sensitivity for the GPS receiver is −120 dBm, but a more realistic level is considered to be around −82 dBm. This value is obtained considering a minimum path loss of 38 dB. This gap is evaluated in [11], calculating the path loss after having generated signals inside the plane. The results show that all the considered values exceed the receiver system sensitivity level but at the same time are under the more realistic value obtained from the path loss. So, the conclusions of the paper are that the radio-frequency emission from the phones tested do not interfere with the avionics system examined, among which is the GPS.

2.3.2 In-Band Signals

Some interference sources broadcast signals whose carrier frequency is allocated in the GNSS bands, and thus they generate in-band interference. Chapter 1 discussed how intersystem and intrasystem interference have to be considered a primary source of in-band disturbance. However, the level of acceptable interference is defined during the design phase of the systems, and, so far, the acceptable level has been the result of international negotiations, discussions, and agreements (consider, e.g., the GPS/Galileo interoperability agreement). In this section the focus is on terrestrial non-GNSS systems. A short description of the most relevant systems emitting power in one or more of the GNSS bands is provided in the following subsections.

Military/Civil Aeronautical Communication Systems

The military communications systems can be considered in-band interferers due to the signal band used by the systems. The Galileo E5a and E5b bands, located within 1164–1214 MHz, occupy frequencies already used for aeronautical radio-navigation services (ARNS) such as for tactical air navigation (TACAN), distance measuring equipment (DME), and secondary surveillance radar (SSR), as well as by the DoD Joint Tactical Information Distribution System (JTIDS) and the Multifunction Information Distribution System (MIDS). Other aeronautical systems operate in these frequencies such as the Traffic Collision and Avoidance System (TCAS), Identification Friend or Foe (IFF), and planned Automatic Dependent Surveillance–Broadcast (ADS-B).

The DME/TACAN systems consist of an airborne interrogator and a ground-based transponder that emits high-power pulsed signals that constitute a threat to GNSS receivers. DME and TACAN provide range measurements of the aircraft with respect to a ground reference station. The TACAN is a military system that provides range and azimuth measurements. The DME/TACAN system operates in the 960- to 1215-MHz frequency band [12] in four different modes: X, Y, W, and Z, even if only the DME/TACAN ground transponder X mode occupies the 1151–1215 MHz frequency band that interferes with the E5a/L5 and E5b GNSS signal (see Figure 2.4 and Table 2.1).

The analytical expression of the classical DME/TACAN pulse pair transmitted by the ground beacons is

$$y_{\text{pulse}}(t) = e^{-\left(\frac{\alpha}{2}\right)t^2} + e^{-\left(\frac{\alpha}{2}\right)(t-\Delta t)^2} \tag{2.1}$$

where, for example, for the X mode $\alpha = 4.5 \cdot 10^{11}\text{s}^{-2}$ and the interpulse interval is $\Delta t = 12\ \mu\text{s}$.

The maximum pulse repetition frequency (PRF) for the DME and the TACAN system are 2700 and 3600 ppps, respectively.

JTID/MIDS are spread-spectrum digital communications systems for exchanging data among military platforms. They operate between 969 and 1206 MHz interfering with the E5a/E5b bands as reported in Figure 2.5.

Ultra-Wideband Signals

The definition of UWB includes any signal that occupies more than 500 MHz between 3.1 and 10.6 GHz and meets the spectrum mask that defines the

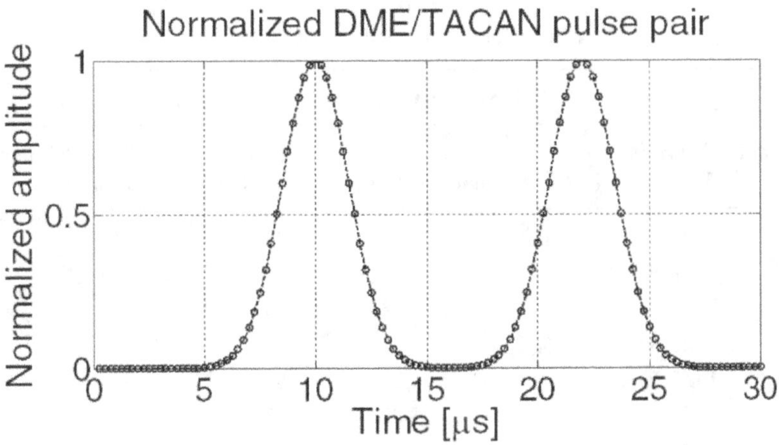

Figure 2.4 Classical baseband DME/TACAN pulse pair.

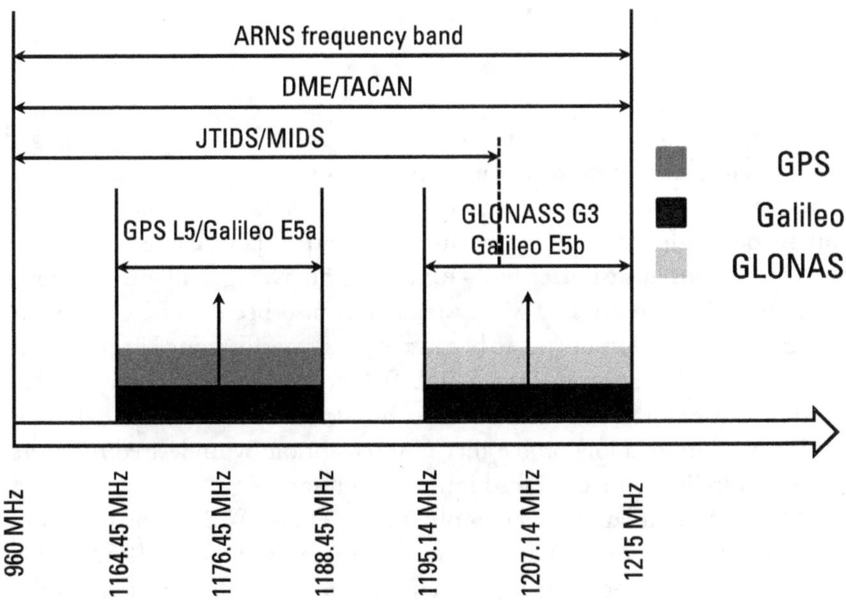

Figure 2.5 DME/TACAN and JTIDS/MIDS frequency plan.

Table 2.1
DME Operational Mode Classification

Channel Mode	Operating Mode	Pulse Pair Spacing (μs) Interrogation	Pulse Pair Spacing (μs) Reply	Time Delay (μs) First Pulse Timing	Time Delay (μs) Second Pulse Timing
X	DME/N	12	12	50	50
	DME/P IA M	12	12	50	—
	DME/P FA M	18	12	56	—
Y	DME/N	36	30	56	50
	DME/P IA M	36	30	56	—
	DME/P FA M	42	30	62	—
W	DME/N	—	—	—	—
	DME/P IA M	24	24	50	—
	DME/P FA M	30	24	56	—
Z	DME/N	—	—	—	—
	DME/P IA M	21	15	56	—
	DME/P FA M	27	15	62	—

indoor limits for UWB communication systems. UWB signals have emerged as a potential solution for low-complexity, low-cost, low-power consumption, and high-data-rate wireless connectivity. The technologies based on UWB offer simultaneous high-data-rate communication, with data transmission rates of 100 to 500 Mbps at distances of 2–10m using an average radiated power of a few hundred microwatts. UWB systems have also been studied for indoor location and navigation purposes because of their performance in multipath environments. The main advantages of UWB are the minimization of reflection from clutter and the ability to penetrate structures with high data rates and high resolution, a low probability of interception by undesired receivers, and the possibility to be used for high-precision ranging.

The data modulation schemes often utilized in UWB systems are pulse position modulation (PPM) and pulse amplitude modulation (PAM). The UWB signal is generated by using sub-nanosecond pulses that spread the signal energy on a wide frequency band. Thus, these systems employ low-power signals but with an extremely wide bandwidth. This aspect is critical for systems such as GNSS whose signal power is far below the noise floor. Several studies

[13–15], concluded that UWB signals can degrade GPS receiver performance. Other studies [16, 17] demonstrate by simulation and in a wireless personal area network (WPAN) case study, respectively, that UWB interference effect can be reduced by choosing the proper modulation parameters.

2.3.3 Classification of Jammers

As explained in Chapter 1, the term *jamming* refers to intentional transmission of RF interference with the goal of masking certain portions of frequency bands with noise. In the case of GNSS systems, a *jammer* (also referred to as a *personal privacy device* (PPD)) is able to jam (or block) GNSS signals, likely preventing the receivers from operating correctly within the jammer area. As an example, Figure 2.6 shows two different models of jammers. Both of them are able to transmit over different frequency bands, including the GNSS E1/L1 band.

Intentional interference is a well-known threat in military applications, but it is also considered a growing concern in the civil environment, thanks

Figure 2.6 Examples of multifrequency GNSS jammers: adjustable desktop jammer (left) and four-antenna portable device (right).

to the fact that real incidents caused by PPDs have already occurred (consider the incident at the Newark Airport in New Jersey, described in Chapter 1).

It is worth recalling that in many countries (e.g., United States or several European countries) jammers are illegal to sell or use. In spite of this, it might not be forbidden to own or buy a jammer, easily achievable through several websites, even at a cost of few tens of dollars [18].

Next, we summarize the main classifications of jammers proposed in the literature and discuss their main characteristics.

A survey of jammers, specifically tailored to in-car jammers, is proposed in [19]. In-car jammers are small devices, powered by a car's cigarette lighter. This class of jammers is particularly important, because their use might allow users (e.g., vehicles) to avoid being tracked.

In [19] jammers are classified into four classes depending on their signal characteristics: A few of them transmit a continuous-wave (CW) signal, while the majority use a chirp signal. The signal bandwidth varies from less than 1 kHz (for the case of CW) to 44.9 MHz, with a sweep time in the interval $[8.62 \div 18.97]$ μs.

A further classification of jammers can be found in [20], where the categories are mainly based on power source. All the jammers analyzed in [20] are portable devices, divided into three groups: devices designed to plug into a car cigarette lighter's 12-volt supply (Group 1), and devices powered by an internal rechargeable battery with (Group 2) or without (Group 3) an external antenna. As a consequence of the analysis of 18 different devices, the authors concluded that all of them use a swept tone method to generate broadband interference on the L1 or L2 band (with a sweep rate of 9 μs, on average, covering a bandwidth of 20 MHz). They also provided an estimate of the analyzed jammers' effective ranges, which vary in the [300m ÷ 6 km] range for tracking, and the [600m ÷ 8.5 km] range for acquisition.

A further survey of jammers can be found in [21], where multifrequency jammers, able to simultaneously disturb more than one GNSS band (L1, L2, and L5), are characterized. The analysis confirmed that cigarette lighter jammers only operate on the L1 band, with different values for the sweep period (with 9 μs being the most common value). It was also shown that the transmitted power can vary from −10 to more than 30 dBm and, in general, cigarette lighter jammers are characterized by lower power levels than multifrequency battery jammers.

An example of a chirp signal generated by a portable jammer device is depicted in Figure 2.7. The figure shows the time-frequency representation of the signal emitted by the jammer. It can be seen that the chirp signal sweeps approximately 9 MHz during an interval of 10 μs.

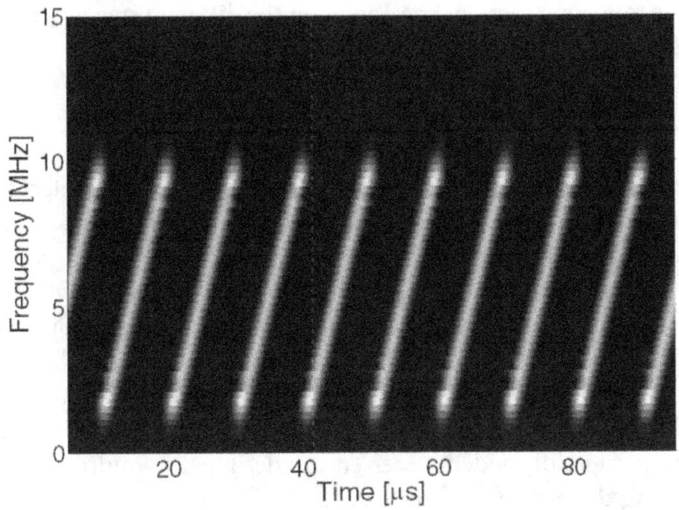

Figure 2.7 Example of chirp signal transmitted by a portable jammer: time–frequency representation.

2.4 The Impact of RFI on GNSS Receivers

When subjected to very strong interference, a GNSS receiver can be totally blinded and stop working. This is often the scope of intentional jammers, who are attempting to deny the use of the GNSS-based positioning in a certain area or region. However, in a number of cases the presence of interference is severe enough to significantly decrease receiver performance, but not severe enough to make the receiver lose its lock on the satellite signals or blind the acquisition of the satellite signals.

Such intermediate power values turn out to be the most dangerous cases, because sometimes they cannot be detected but they are leading to a worsening of the positioning performance. For the user of a GNSS receiver, the relevance of the effect of strong RFI is obvious. If the receiver is unable to track satellites, it cannot calculate its position. When the receiver is able to track satellites, but the signal is affected by RFI, the error on the pseudorange measurements is increased. The accuracy of the position solution depends, among others, on the quality of the pseudorange measurements and/or the phase measurements. As a consequence, when RFI degrades the pseudorange and phase measurements or induces cycle slips on the phase measurements, the accuracy of the position solution will decrease.

In the following sections the effects on the different stages of the receiver are discussed.

2.4.1 Impact on the Front-End

The front-end of the receiver is the first stage of the receiver affected by the presence of an interfering source. The front-end filters the incoming signal in the bandwidth of interest, demodulating it to the chosen intermediate frequency before performing the analog-to-digital conversion (ADC).

We must consider the presence in the front-end of the adjustable gain control (AGC) between the analog portion of the front-end and the ADC. The variable gain amplifier adjusts the power of the incoming signal to optimize the signal dynamics for the ADC in order to minimize quantization losses. Nowadays, in fact, all modern receivers are designed as multibit equipment, thus requiring the presence of an AGC.

When the GNSS band is interference free, which should be the norm due to restrictions on emissions in and near the band, AGC gain depends almost exclusively on thermal noise, since the received GNSS signal power level is below that of the thermal noise floor. The primary role of the AGC is to adjust the signal dynamics for variations in the received power due to the elevation of the satellite and/or different active antenna gain values.

The statistics for samples at the ADC output in the case of an interference-free GNSS band, reported in Figure 2.8(a), are basically normally distributed, as shown in Figure 2.8(b).

When in-band interference is present, the AGC will squeeze the incoming signal in order to match the maximum dynamics of the ADC, thus causing a reduction of the amplitude of the useful signal, which may be lost. This is typical of what may happen in the presence of some kind of WBI; that being spread over a bandwidth larger than the passband of the front-end filter can be seen as additional noise on the bandwidth of interest.

Furthermore, in the presence of NBI or CWI, the statistics of the digital signal at the output of the ADC are also affected. This can be seen in Figure 2.8(d), where the boundary quantization levels become more probable than the others. In this case the AGC is still able to compress the input signal to avoid a stronger saturation, however, the following stages of the receiver will have to deal with a GNSS contribution that is quantized only on the lower levels.

In the presence of stronger interference, even the other components of the front-end (filters and amplifiers) may be led to work outside of their nominal regions, generating nonlinear effects or clipping phenomena (in which the signal amplitude exceeds the hardware's capability to treat them). In both cases spurious harmonics are generated and mixed with the useful signal in the front-end itself.

2.4.2 Impact on the Acquisition Stage

If the interference is not driving the AGC/ADC to full saturation, the acquisition module is still able to perform its task, processing the interfered signal to estimate the code phase and the Doppler shift with respect to the local code.

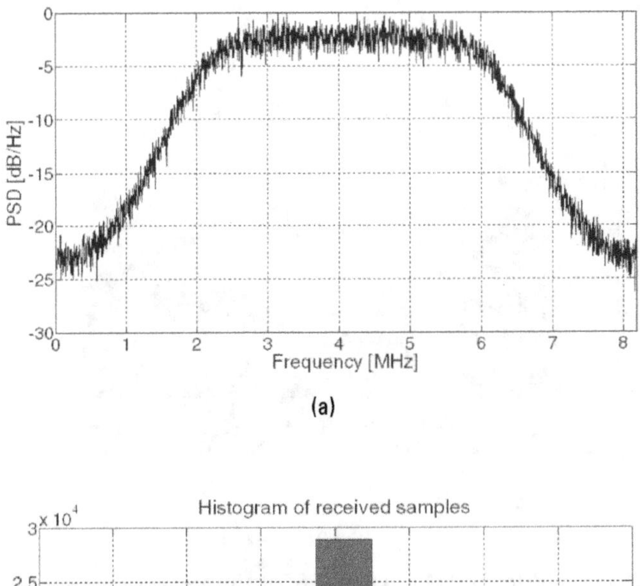

(a)

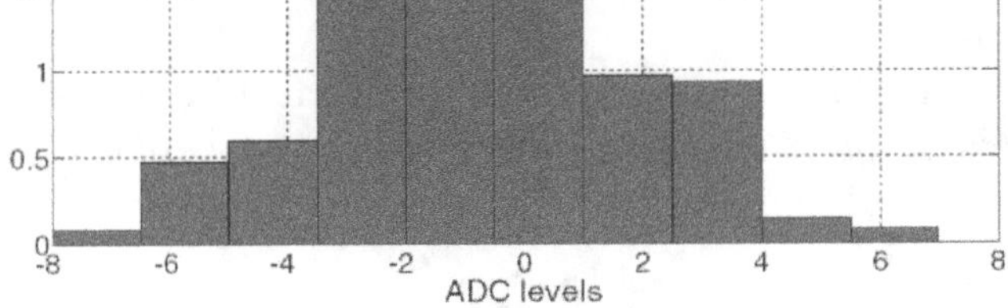

(b)

Figure 2.8 (a) GPS L1 C/A code PSD in the absence of interference. (b) Histograms of the samples at the ADC output in the absence of interference.

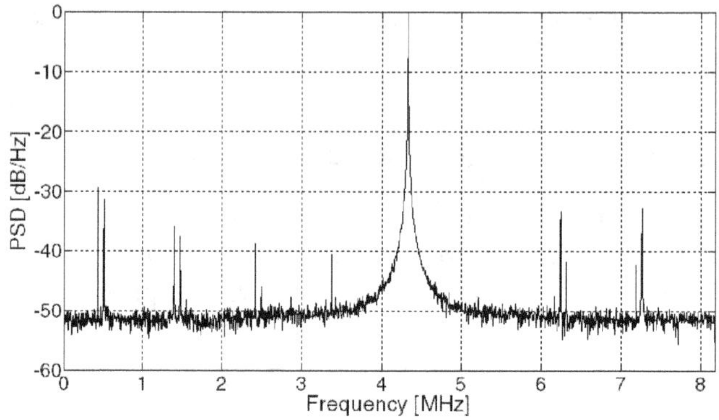

(c)

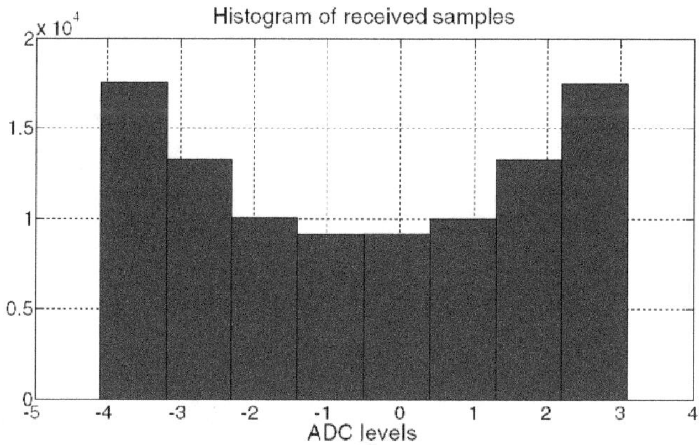

(d)

Figure 2.8 *(Continued)* (c) GPS L1 C/A code PSD in the presence of CWI. (d) Histograms of the samples at the ADC output in presence of CWI.

The correlation with the local code can be seen as a spreading operation followed by a filter.

In [22] an exhaustive derivation of the impact of CWI on the acquisition stage of a GNSS receiver is provided. In the presence of CWI the expression of the digitized signal[2] at the input of the baseband processing block of a GNSS receiver is

[2] For the sake of simplicity, in this description the effect of the data and of the quantization effects are neglected.

$$y_{\mathrm{IF}}[n] = \sqrt{2C}c\left[n - \tau_0\right]\cos\left(2\pi\left(f_{\mathrm{IF}} + f_{D,0}\right)T_s n + \varphi_0\right)$$
$$+A_{\mathrm{int}}\cos\left(2\pi F_{\mathrm{int}} T_s n + \theta_{\mathrm{int}}\right) + \eta[n] \quad (2.2)$$

where the first member of the sum is the useful received GNSS signal; A_{int}, F_{int}, and θ_{int} are, respectively, the amplitude, the carrier frequency, and the random phase uniformly distributed in the range $[-\pi ; \pi)$ of the CWI assumed to be a pure sinusoid; and $W_{\mathrm{IF}}[n]$ is the Gaussian noise component that, under the Nyquist sampling theorem assumption, can be assumed to be a classical independent and identically distributed (IID) discrete random process.

According to the equivalent scheme of a GNSS acquisition block depicted in Figure 2.9, the signal in (2.2) is first multiplied by a complex exponential and then multiplied and integrated with respect to the local code chosen according to an hypothesis of Doppler frequency f_D and code delay τ, thus providing the following cross ambiguity function complex components:

$$S_I\left(\tau, f_D\right) = \frac{1}{N}\sum_{n=0}^{N-1} r_I[n]c[n-\tau] = r_I[\tau] * h_c[\tau]$$
$$S_Q\left(\tau, f_D\right) = \frac{1}{N}\sum_{n=0}^{N-1} r_Q[n]c[n-\tau] = r_Q[\tau] * h_c[\tau] \quad (2.3)$$

where h_c is the equivalent filter representing the operation of multiplication and integration with respect the local code, and N is the number of coherent integrated samples. Finally the CAF is obtained as a complex modulus

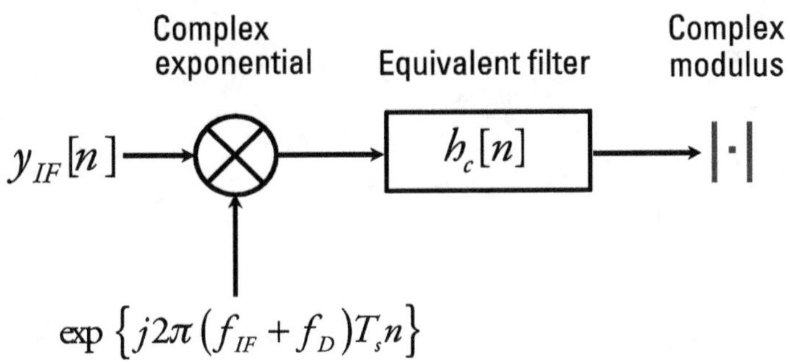

Figure 2.9 Equivalent scheme of a GNSS acquisition block.

$$S(\tau, f_D) = \sqrt{S_I(\tau, f_D)^2 + S_Q(\tau, f_D)^2} \tag{2.4}$$

When the code delay and Doppler shift are correctly recovered, the useful signal contribution is given by

$$S_y \approx \sqrt{C/2} \cdot \exp\{j\varphi_0\} \tag{2.5}$$

As derived in [22], the CWI contribution to the CAF is composed of two complex exponentials at different frequencies generated by the multiplication of the CWI with the complex exponential:

$$\begin{aligned} I_D[n] &= \frac{A_{\text{int}}}{2}\exp\left\{j2\pi\left(F_{\text{int}} + (f_{\text{IF}} + f_{D,0})\right)T_s n + j\theta_{\text{int}}\right\} \\ &+ \frac{A_{\text{int}}}{2}\exp\left\{-j2\pi\left(F_{\text{int}} - (f_{\text{IF}} + f_{D,0})\right)T_s n - j\theta_{\text{int}}\right\} \end{aligned} \tag{2.6}$$

Thus, these two components are fed to the equivalent filter $h_c[n]$ whose output is

$$\begin{aligned} S_{\text{int}} &= k_1\frac{A_{\text{int}}}{2}\exp\left\{j2\pi\left(F_{\text{int}} + \left(f_{\text{IF}} + f_{D,0}\right)\right)T_s\tau_0 + j\theta_{\text{int}} + j\theta_1\right\} \\ &+ k_2\frac{A_{\text{int}}}{2}\exp\left\{-j2\pi\left(F_{\text{int}} - \left(f_{\text{IF}} + f_{D,0}\right)\right)T_s\tau_0 - j\theta_{\text{int}} + j\theta_2\right\} \end{aligned} \tag{2.7}$$

where

$$\begin{aligned} k_i^2 &= \left|H_c\left(\pm F_{\text{int}} + f_{\text{IF}} + f_{D,0}\right)\right|^2 = \int_{-\infty}^{\infty}\left|H_c(f)\right|^2\delta\left(f - \left(\pm F_{\text{int}} + f_{\text{IF}} + f_{D,0}\right)\right)df \\ &= \int_{-\infty}^{\infty} G_s(f)G_i(f)\,df \end{aligned} \tag{2.8}$$

with $G_s(f) = \left|H_c\left(f + f_{\text{IF}} + f_{D,0}\right)\right|^2$, $G_i = \delta(f \pm F_{\text{int}})$, and $\delta(\cdot)$ is the delta Dirac. Thus, k_1^2 and k_2^2 are nothing but spectral separation coefficients according to (2.8).

Concerning the noise contribution, the multiplication with the complex exponential splits the noise power on the two acquisition branches (in-phase and quadrature). As stated in [22], the total variance of the output process is given by

$$\sigma_{\text{out}}^2 = \frac{1}{N}\sigma_{\text{IF}}^2 = \frac{1}{N}N_0\beta_r \tag{2.9}$$

where $N_0/2$ is the power spectral density (PSD) of the IF noise and β_r is the IF filter bandwidth. Thus, the noise contribution to the CAF can be assumed to have a Gaussian distribution with zero mean and variance for the in-phase and quadrature component equal to $\sigma_{\text{out}}^2/2$ ($S_W \sim N_c\left(0, \frac{\sigma_{\text{out}}^2}{2}\mathbf{I}_2\right)$ with $\mathbf{I}_2$ being a 2×2 identity matrix).

In conclusion, in the presence of CWI and assuming perfect recovery of Doppler frequency and code delay of the received GNSS signal, the CAF becomes the following Rice distribution:

$$S(\tau, f_D)|\varphi_0, \theta_{\text{int}} = \frac{x}{\sigma^2}\exp\left\{-\frac{x^2+\alpha^2}{2\sigma^2}\right\}I_0\left(\frac{x\alpha}{\sigma^2}\right), \quad x > 0 \tag{2.10}$$

where $\alpha^2 = |S_y + S_{\text{int}}|^2$, $\sigma^2 = \sigma_{\text{out}}^2/2$ and I_0 is the modified Bessel function of the first kind and zero order. Thus, it is possible to derive a detection probability conditioned with the starting knowledge of φ_0 and θ_{int} as

$$P_d(\beta|\varphi_0, \theta_{\text{int}}) = \int_{\beta}^{\infty}\frac{x}{\sigma^2}\exp\left\{-\frac{x^2+\alpha^2}{2\sigma^2}\right\}I_0\left(\frac{x\alpha}{\sigma^2}\right)dx = Q\left(\frac{\alpha}{\sigma};\frac{\beta}{\sigma}\right) \tag{2.11}$$

Equation (2.11) assumes precise knowledge of the initial phase of the GNSS signal and the interference, which in general are unknowns. In fact, the impact of the CWI on the search space changes depending on such parameters; thus, as mentioned in [22], the overall detection probability has to be computed by removing the hypothesis of the initial knowledge of φ_0 and θ_{int}, and averaging with respect to the probability density function (PDF) of these two random variables. More details on acquisition performance in the presence of CWI can be found in [22].

Figure 2.10 shows the acquisition search space for different levels of the interfering power of a CWI from −140 to −130 dBW compared to the interference-free case. The search spaces depicted in Figure 2.10 for the four scenarios are achieved using 1 ms of coherent integration time and three non-coherent accumulations, and the peak-to-noise-floor separation defined as

$$\alpha_{\text{mean}} = \frac{R_p}{M_c} \tag{2.12}$$

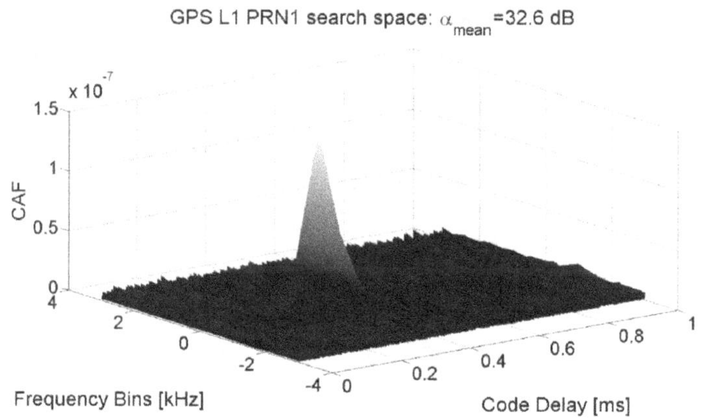

(a)

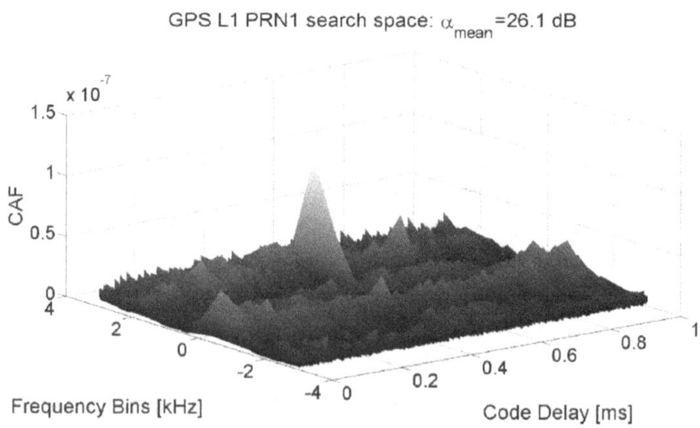

(b)

Figure 2.10 GPS L1 C/A acquisition search space in (a) an interference-free environment and in the presence of (b) –140 dBW in-band CWI.

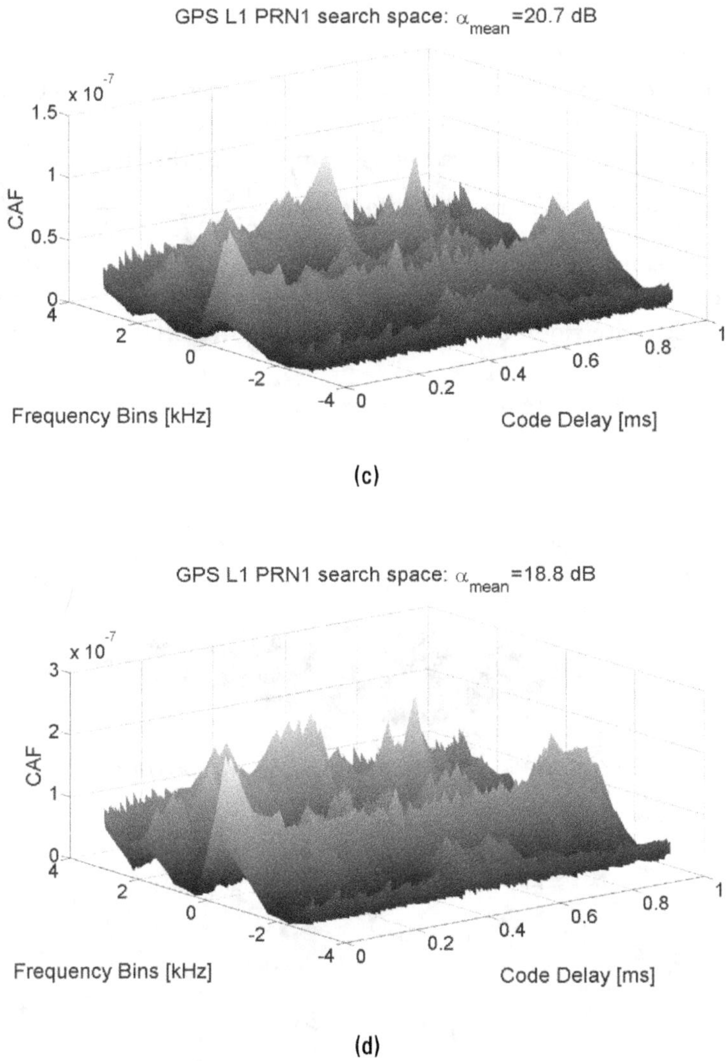

Figure 2.10 *(Continued)* (c) −135 dBW in-band CWI, and (d) −130 dBW in-band CWI.

is considered as a figure of merit. The value of α_{mean} decreases as the interfering power increases, thus increasing the probability of a false alarm. Furthermore, with the increasing power of the CWI, a modulation effect in the search space floor in the Doppler domain dimension can be observed. Such an effect is mainly determined by the new harmonics components generated by the multiplication between the locally generated carrier and received CWI. Such an effect also depends on how the interfering signal and the useful GNSS signal are combined at the entrance to the acquisition block, which in turn depends on the random variables φ_0 and θ_{int}.

When in the presence of WBI, a different effect is observed in the acquisition search space. Considering a band-limited Gaussian white noise spread all over the GNSS useful filtered signal components, the effect on the CAF envelope is an increase in the noise floor from $\frac{1}{N}N_0\beta_r$ to $\frac{1}{N}(N_0 + I_0)\beta_r$, with $I_0/2$ being the PSD of the additional band-limited noise at IF. Thus, the effect on the acquisition search space is represented by an increase in the search space noise floor, as demonstrated in Figure 2.11. The presence of

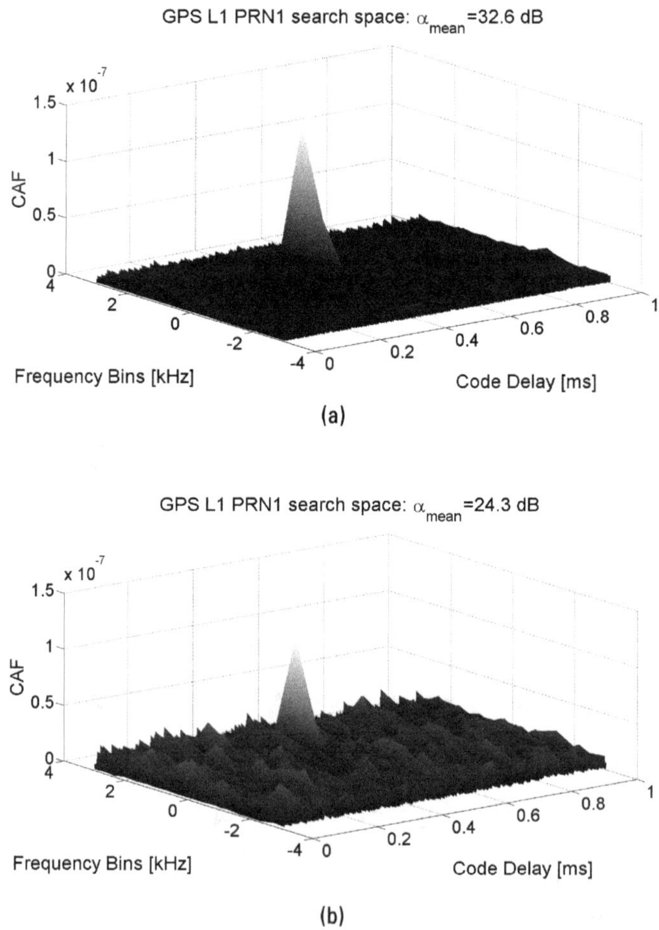

Figure 2.11 GPS L1 C/A acquisition search space in (a) an interference-free environment and in the presence of (b) −140 dBW NBI.

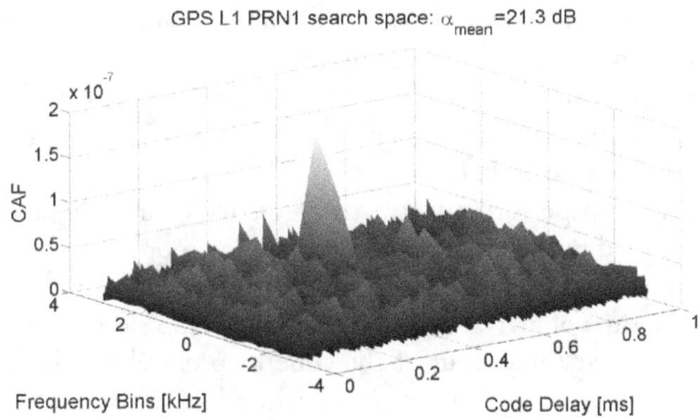

(c)

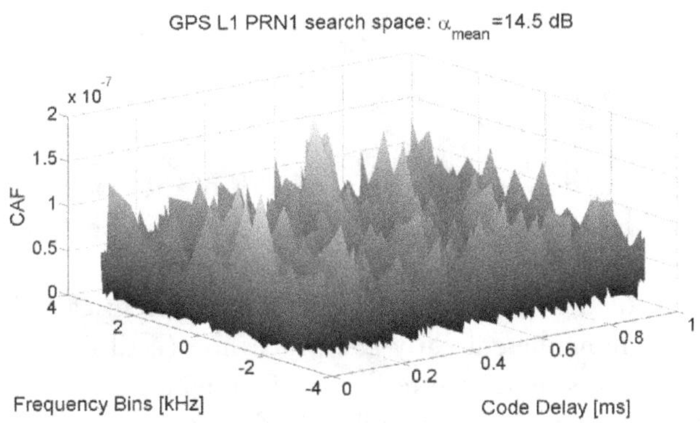

(d)

Figure 2.11 *(Continued)* (c) −135 dBW NBI, and (d) −130 dBW NBI.

additive band-limited noise causes a uniform increase in the noise floor in the search space that might mask the correct correlation peak and thus fool the acquisition process, as in the case shown in Figure 2.11(d).

An extensive study of the effects of several types of interference on the acquisition probabilities can be found in [23].

2.4.3 Impact on the Tracking Stage

The impact of the interferer on the tracking stage has a direct consequence on the quality of the measured pseudorange. The presence of harmful interfering signals not only produces an increase in the variance of the time-of-arrival

(TOA) estimate provided by the discriminator but also causes a modification of the shape of the S-curve of the code discriminator, thus creating in some cases a bias in the measurements [23]. A method for assessing the bias on the discriminator function caused by the presence of CWI and narrowband interference is provided in [24].

Tracking loop performance in the presence of interference has been widely investigated in the literature in terms of the root-mean-square (RMS) code tracking error. The code tracking loop is in charge of providing a fine estimate of the time of arrival τ of the received GNSS signal; thus, the variance of the signal at the output of the code loop correlator directly affects the quality of the pseudorange measurements. In [25] the variance of the smoothed TOA estimate σ_s^2, provided by the code tracking loop, is derived as a function of the unsmoothed TOA estimate σ_u^2, which commonly comes from the discriminator output, as

$$\sigma_s^2 \cong \sigma_u^2 2B_L T\left(1-0.5B_L T\right) \tag{2.13}$$

where B_L is the one-side equivalent bandwidth of the tracking loop and T is the employed integration time. Equation (2.13) can be approximated as $\sigma_s^2 \cong \sigma_u^2 2B_L T$ for small values of $B_L T$.

A very detailed derivation of σ_u^2 for both coherent early-late processing (CELP) and noncoherent early-late processing (NELP) in the presence of narrowband and wideband interference is provided in [25] and [26]. In the presence of a coherent early-late discriminator, the variance of the code tracking error is

$$\sigma_{s,\mathrm{CELP}}^2 = \frac{B_L\left(1-0.5B_L T\right)\int_{-\beta_r/2}^{\beta_r/2} G_w(f)G_s(f)\sin^2(\pi f\Delta)\,df}{(2\pi)^2 C\left(\int_{-\beta_r/2}^{\beta_r/2} fG_s(f)\sin(\pi f\Delta)\,df\right)^2} \tag{2.14}$$

where:

β_r = the two-sided front-end filter bandwidth;

Δ = the early-late spacing (in unit of seconds);

$G_s(f)$ = the GNSS signal power spectral density normalized to unit power over infinite bandwidth $\int_{-\infty}^{\infty} G_I(f)\,df = 1$;

C = the GNSS received signal carrier power;

$G_w(f) = N_0 + C_I G_I(f)$ = the noise plus interference power spectral density, with N_0 being the flat noise power spectral density over the receiver front-end bandwidth, C_I the interference carrier power over an infinite bandwidth, and $G_I(f)$ the interference power spectral density normalized such that $\int_{-\infty}^{\infty} G_I(f)\,df = 1$.

Splitting the noise and the interference component, (2.14) can be rewritten as

$$\sigma^2_{s,\text{CELP}} = \frac{B_L\left(1-0.5B_LT\right)\int_{-\beta_r/2}^{\beta_r/2} G_s(f)\sin^2(\pi f\Delta)\,df}{(2\pi)^2\frac{C}{N_0}\left(\int_{-\beta_r/2}^{\beta_r/2} fG_s(f)\sin(\pi f\Delta)\,df\right)^2} + \frac{B_L\left(1-0.5B_LT\right)\int_{-\beta_r/2}^{\beta_r/2} G_I(f)G_s(f)\sin^2(\pi f\Delta)\,df}{(2\pi)^2\frac{C}{C_I}\left(\int_{-\beta_r/2}^{\beta_r/2} fG_s(f)\sin(\pi f\Delta)\,df\right)^2} \tag{2.15}$$

According to the theoretical derivation proposed in [26], for noncoherent early-late processing, the variance of the code tracking error becomes

$$\sigma^2_{s,\text{NELP}} = \sigma^2_{s,\text{CELP}}\left[1+\frac{\int_{-\beta_r/2}^{\beta_r/2} G_s(f)\cos^2(\pi f\Delta)\,df}{T\frac{C}{N_0}\left(\int_{-\beta_r/2}^{\beta_r/2} G_s(f)\cos(\pi f\Delta)\,df\right)^2} + \frac{\int_{-\beta_r/2}^{\beta_r/2} G_I(f)G_s(f)\cos^2(\pi f\Delta)\,df}{T\frac{C}{C_I}\left(\int_{-\beta_r/2}^{\beta_r/2} G_s(f)\cos(\pi f\Delta)\,df\right)^2}\right] \tag{2.16}$$

Figure 2.12 shows a comparison between the CELP and NELP RMS code tracking error of GPS L1 C/A code in the presence of additive filtered Gaussian noise around the intermediate frequency and with a 200-kHz bandwidth, as a function of the interference to signal power ratio C_I/C.

As expected, an increase in the interference power leads to an increase in the code tracking accuracy. Furthermore, note that for low values of C_I/C,

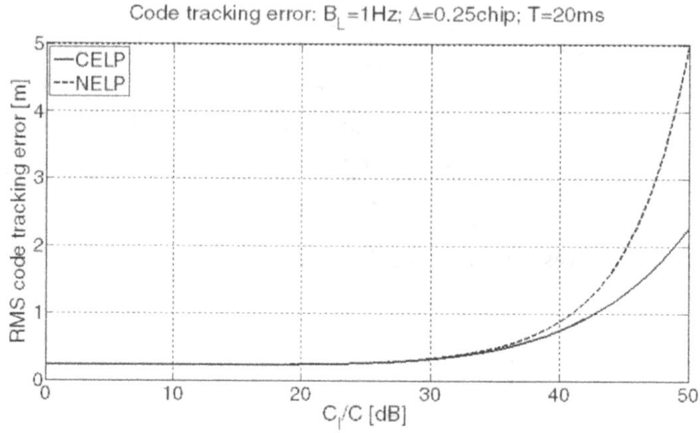

Figure 2.12 GPS L1 C/A code tracking error comparison: CELP versus NELP.

CELP and NELP show similar performance, whereas for high values of C_I/C, NELP code tracking performance is worse than CELP code tracking performance. This is mainly due to the fact that for high interfering power values, and thus low signal-to-noise plus interference ratios (SNIRs), the effect of the so called "squaring loss" is predominant.

According to the model in (2.16), an analysis of the impact of filtered Gaussian white noise characteristics on code tracking accuracy is provided for both BPSK(1) and BOC(1,1) modulation in Figure 2.13. For both code tracking error analyses, a front-end bandwidth β_r equal to 4.092 MHz and a sampling frequency f_s equal to 16.36 MHz are considered. Furthermore, a nominal carrier–to–noise density ratio C/N_0 equal to 47 decibels per hertz (dB-Hz) is assumed as well as fixed interference for a signal power ratio C_I/C equal to 40 dB.

Figures 2.13(a) and (b) show, respectively, the tracking code accuracy of BPSK(1) and BOC(1,1) signals according to (2.16) in the presence of filtered Gaussian white interference with bandwidth sweeping from 50 kHz up to 4.092 MHz centered on the intermediate frequency. In both figures, note that the code tracking error is degraded for wider early-late spacing. Lower values of spacing are less sensitive to the increase of the interference bandwidth. In the case of BPSK(1) tracking code accuracy, the error is larger for an interference bandwidth close to the chipping rate, whereas for BOC(1,1) modulation, an interference bandwidth approximately twice the chipping rate wider leads to the highest code tracking error. A more detailed analysis on band-limited white interference is provided in [26].

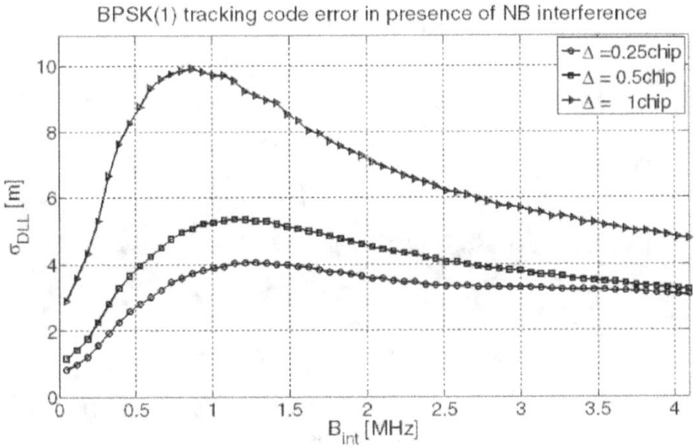

(a)

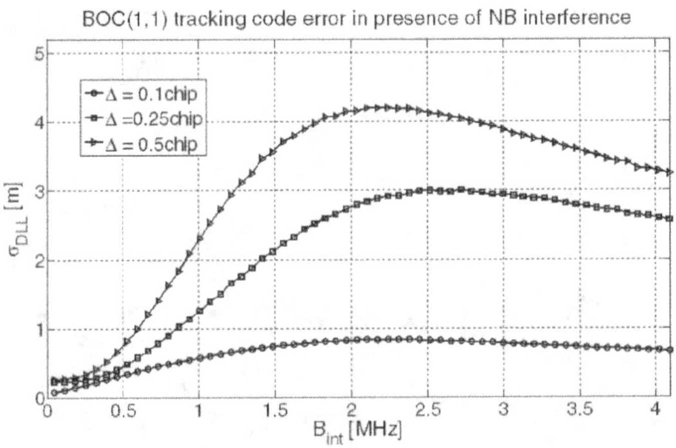

(b)

Figure 2.13 Tracking code accuracy of (a) BPSK(1) and (b) BOC(1,1) signals in the presence of NBI interference.

As an example of the effect of interference on the tracking stage, Figure 2.14 depicts the outputs of the early-prompt-late correlators. In the presence of in-band CWI and of NBI, the interference is injected 9.3s after the beginning of the tracking stage where the receiver is correctly locked on the received GNSS signal. Table 2.2 provides a summary of the tracking parameter configuration adopted in the GNSS receiver for such an example.

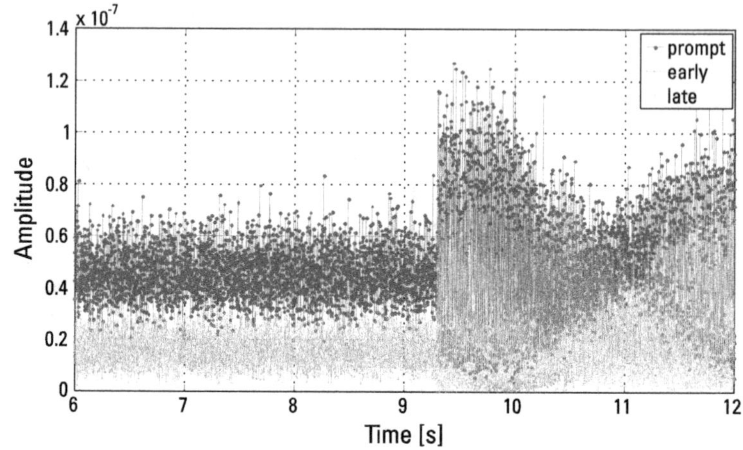

(a)

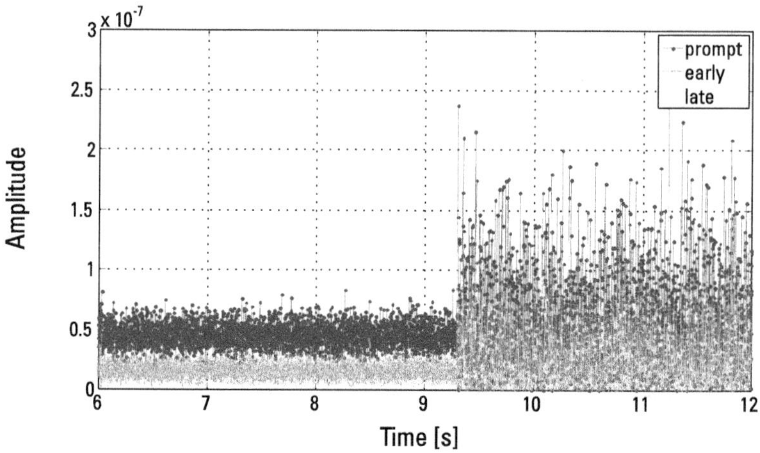

(b)

Figure 2.14 GPS L1 C/A tracking performance: early-prompt-late correlators in the presence of (ab) –130 dBW in-band CWI and (b) –130 dBW NBI.

Table 2.2
GNSS Receiver Tracking Loop Parameters Configuration

Predetection Integration Time	**PLL B_n**	**DLL B_n**	**Spacing Δ**
1 ms	10 Hz	2 Hz	0.9 chip

The presence of a CWI, shifted 200 kHz with respect the GNSS signal intermediate frequency (thus in correspondence with a C/A code spectrum line), not only increases the noise at the correlators outputs but leads to harmonic behavior on the part of the early-prompt-late correlator outputs.

The presence of NBI increases the variance of the correlators' outputs and this has a direct consequence on the increase of the pseudorange error. The effect of the increased pseudorange measurement in presence of CWI and NBI is proved by the trend of the discriminator outputs shown in Figure 2.15.

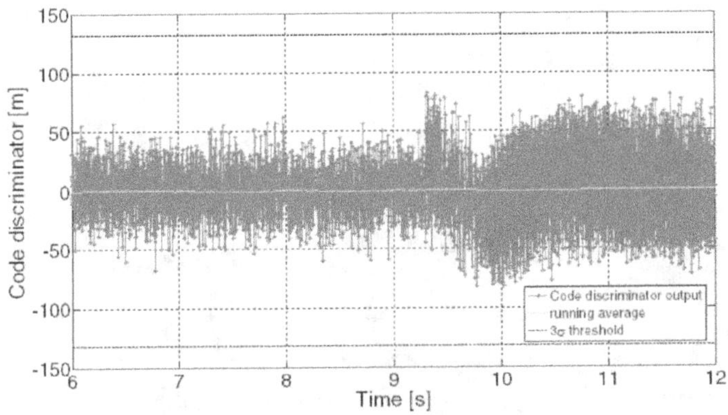

(a)

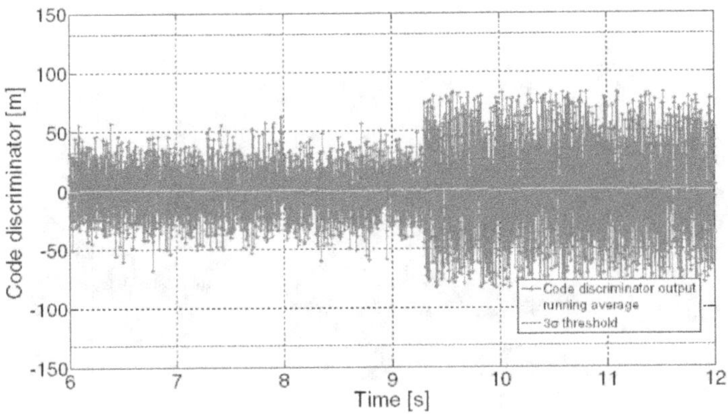

(b)

Figure 2.15 GPS L1 C/A tracking performance: code discriminator output in the presence of (a) −130 dBW in-band CWI and (b) −130 dBW NBI.

The noise on the phase measurements of the GNSS receiver is increased as well. Figures 2.16(a) and (b) show, respectively, the different effect on the carrier phase discriminator output caused by the presence of in-band CWI and band-limited NBI. The presence of additive band-limited noise leads to an overall increase in the carrier phase discriminator output variance over the 3σ threshold, which for a PLL two-quadrant arctangent discriminator is 45° [see Figure 2.16(b)]. When in presence of a strong CWI, a sudden jump of the phase discriminator output is detected as soon as the CWI is injected onto the received signal, as shown in Figure 2.16(a).

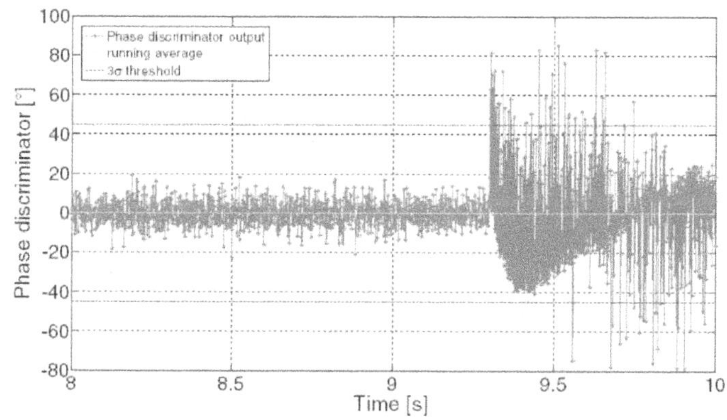

(a)

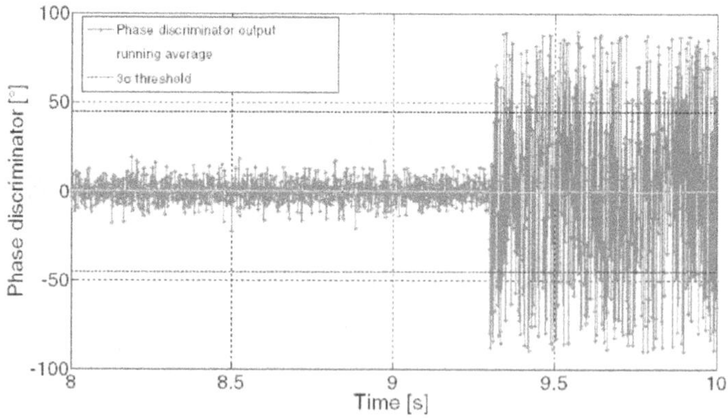

(b)

Figure 2.16 GPS L1 C/A tracking performance: carrier phase discriminator output in the presence of (a) −120 dBW in-band CWI and −120 dBW NBI.

2.4.4 Impact on the Estimated Signal-to-Noise Ratio

Interference affects the estimated signal-to-noise ratio (C/N_0 in decibels per hertz (dB-Hz)), as shown in Figure 2.17, where the trend of the C/N_0 is reported for CWI in part (a) and NBI in part (b) for different interference power levels. Note that sticking to the definition of C/N_0 as the ratio between the received power and the power spectral density due to thermal noise at the input of

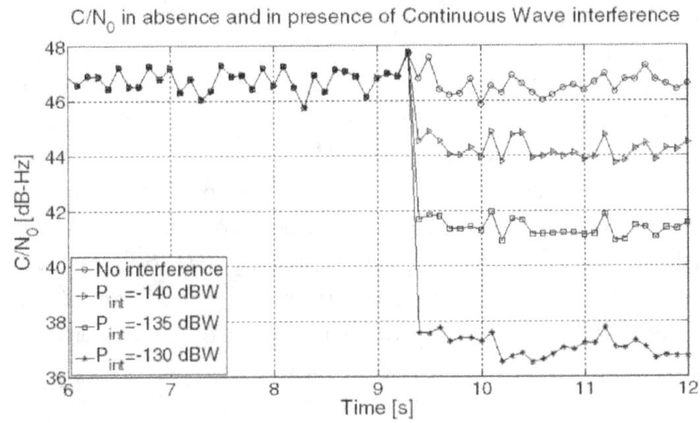

(a)

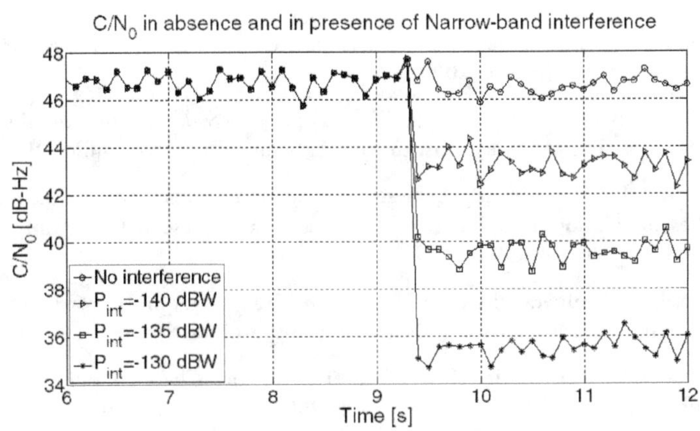

(b)

Figure 2.17 Carrier–to–noise density ratio estimation in the presence of (a) CWI and (b) NBI.

the receiver, the presence of interference should not change the value, since the thermal noise is not increasing. However, the C/N_0 value provided by the GNSS receivers is estimated on the basis of the correlator outputs at the tracking stage. For this reason the estimation is affected by the presence of the additional (nonthermal) noise generated by the interference. The variation of C/N_0 can also be used as observable for interference (or other threats) detection as explained in Chapter 5.

2.5 Conclusions

This chapter provided an overview of the main terrestrial sources potentially generating RFI that can affect GNSS systems. Despite the large number of possible threats, it should be noted that RFI is typically generated only in the case of a poorly designed communication systems or during malfunctioning events. Furthermore, spurious emissions are also attenuated with the distance from the transmitter, thus making it a threat only for GNSS receivers operating in a close area. Nevertheless, as shown in the second part of the chapter, RFI can affect all stages of the processing chain of a receiver, leading to a worsening of the position information it provides.

References

[1] Kaplan, E. D., and C. Hegarty, *Understanding GPS: Principles and Applications*, Norwood, MA: Artech House, 2005.

[2] Parkinson, B. W., and J. J. Spilker, *Global Positioning System: Theory and Applications*, Washington, DC: American Institute of Aeronautics and Astronautics, 1996.

[3] Landry, R. J., and A. Renard, "Analysis of Potential Interference Sources and Assessment of Present Solutions for GPS/GNSS Receivers," paper presented 4th Saint-Petersburg on INS, May 26–28, 1997.

[4] Volpe, J. A., "Vulnerability Assessment of the Transportation Infrastructure Relying on the Global Position System," National Transportation Systems, January 2000.

[5] "Digital Video Broadcasting (DVB): Framing Structure, Channel Coding and Modulation for Digital Terrestrial Television, 2004–2006," Sophia-Antipolis, France: European Telecommunications Standards Institute.

[6] Buck, T., and G. Sellick, "GPS RF Interference via a TV Video Signal," *Proc. 10th Int. Technical Meeting of the Satellite Division of the Institute of Navigation (ION GPS 1997)*, Kansas City, MO, September 1997, pp. 1497–1501.

[7] Motella, B., M. Pini, and F. Dovis, "Investigation on the Effect of Strong Out-of-Band Signals on Global Navigation Satellite Systems Receivers," *GPS Solutions*, Vol. 12, No. 2, March 2008, pp. 77–86.

[8] Borio, D., S. Savasta, and L. Lo Presti, "On the DVB-T Coexistence with Galileo and GPS Systems," *Proc. 3rd ESA Workshop on Satellite Navigation User Equipment Technologies (NAVITEC 2006)*, ESA/ESTEC, Noordwijk, The Netherlands, December 2006.

[9] Dimos, G., T. Upadhyay, and T. Jenkins, "Low Cost Solution to Narrowband GPS Interference Problem," *Proc. NAECON*, 1995.

[10] Kuriger, G., et al., "Investigation of Spurious Emission from Cellular Phones and the Possible Effect on Aircraft Navigation Equipment," *IEEE Trans. on Electromagnetic Compatibility*, Vol. 45, No. 2, 2003, pp. 281–292.

[11] RTCA, "Portable Electronic Devices Carried on Board Aircraft," 1997; available at http//www.rtca.org.

[12] Bastide, F., et al., "GPS L5 and Galileo E5a/E5b Signal-To-Noise Density Ratio Degradation Due to DME/TACAN Signals: Simulations and Theoretical Derivation," *Proc 2004 National Technical Meeting of the Institute of Navigation*, San Diego, CA, January 2004, pp. 1049–1062.

[13] Hamalainen, M., et al., "On the UWB System Coexistence with GSM900, UMTS/WCDMA, and GPS," *IEEE J. on Selected Areas in Communications*, Vol. 20, No. 9, December 2002, pp. 1712–1721.

[14] Cummings, D. A., "Aggregate Ultra Wideband Impact on Global Positioning System Receivers," *Proc. Radio and Wireless Conference (RAWCON 2001)*. pp.101–104, 2001. doi:10.1109/RAWCON.2001.947539

[15] Andenon, D. S., et al., "Assessment of Compatibility Between Ultrawideband Systems and Global Positioning System (GPS) Receivers," NT1A Special Publication, February 2001.

[16] Morton, Y. T., et al., "A Software Approach to Access Ultra-Wide Band Interference on GPS Receivers," *Proc. Position Location and Navigation Symposium (PLANS 2004)*, pp. 551–557, April 26–29, 2004. doi:10.1109/PLANS.2004.1309041

[17] Giuliano, R., and F. Mazzenga, "On the Coexistence of Power-Controlled Ultrawide-Band Systems with UMTS, GPS, DCS1800, and Fixed Wireless Systems," *IEEE Trans. on Vehicular Technology*, Vol. 54, No. 1, pp. 62–81, January 2005. doi:10.1109/TVT.2004.838843

[18] Pullen, S., and G. X. Gao, "GNSS Jamming in the Name of Privacy," *Inside GNSS*, Vol. 7, No. 2, March/April 2012, pp. 34–43.

[19] Kraus, T., R. Bauernfeind, and B. Eissfeller, "Survey of In-Car Jammers—Analysis and Modeling of the RF Signals and IF Samples (Suitable for Active Signal Cancelation)," *Proc 24th Int. Technical Meeting of the Satellite Division of the Institute of Navigation (ION GNSS 2011)*, Portland, OR, September 2011, pp. 430–435.

[20] Mitch, R. H., et al., "Know Your Enemy: Signal Characteristics of Civil GPS Jammers," *GPS World*, January 2012, Vol. 24, No. 1, pp. 64–71.

[21] Borio, D., J. Fortuny-Guasch, and C. O'Driscoll, "Characterization of GNSS Jammers," *Coordinates*, Vol. IX, No. 5, May 2013, pp. 8–16.

[22] Borio, D., "GNSS Acquisition in the Presence of Continuous Wave Interference," *IEEE Trans. on Aerospace and Electronic Systems*, Vol. 46, No. 1, January 2010, pp. 47–60.

[23] Wildemeersch, M., et al., "Impact Study of Unintentional Interference on GNSS Receivers," European Commission Joint Research Center. doi:10.2788/57794

[24] Motella, B., et al., "Method for Assessing the Interference Impact on GNSS Receivers," *IEEE Trans. on Aerospace and Electronic Systems*, Vol. 47, No. 2, 2011, pp.1416–1432.

[25] Betz, J. W., and K. R. Kolodziejski, "Generalized Theory of Code Tracking with an Early-Late Discriminator Part I: Lower Bound and Coherent Processing," *IEEE Trans. on Aerospace and Electronic Systems*, Vol. 45, No. 4, October 2009, pp. 1538–1556.

[26] Betz, J. W., and K. R. Kolodziejski, "Generalized Theory of Code Tracking with an Early-Late Discriminator Part II: Noncoherent Processing and Numerical Results," *IEEE Trans. on Aerospace and Electronic Systems*, Vol. 45, No. 4, October 2009, pp. 1557–1564.

3

The Spoofing Menace

Davide Margaria and Marco Pini

3.1 Introduction: Meaconing and Spoofing Attacks

The presence of GNSS technology in the modern lifestyle has been constantly growing in the last years, supporting the use of GNSS receivers in a large variety of applications. In this context, the mass market sector drives the use of navigation technologies in new and emerging applications, thanks in part to the widespread use of smartphones and tablets with embedded GNSS chipsets [1]. Examples of applications are road tolling, pay as you drive, location-based services, communication network synchronization, financial transactions, transportation, and fleet management.

GNSS receivers are vulnerable to intentional interference and this creates incentives for attackers who want to impair or fool systems based on GNSS. This issue could have a serious impact in many applications, because security is not a "built-in feature of GNSS open service," as highlighted in [2]. In fact, at least according to the current version of the interface control documents (ICDs), neither the GPS L1 C/A code [3] nor the E1 Galileo OS signal [4] implements any means in the receivers to ensure the authenticity of the source of the data (i.e., signal authentication) or to improve the robustness of the receiver against possible attacks (i.e., antijamming and/or antispoofing). The same remarks are also applicable to current civil GLONASS signals [5].

In terms of the potential signal vulnerability of mass market GNSS receivers, the sheer number of applications based on these receivers is in some ways focusing attention on the need to have some form of assurance regarding the receivers' outputs (both time and position). Recently, many works available in the literature [2, 6–8] have highlighted the risk of intentional attackers willing to disrupt the GNSS receiver functions.

From a general perspective, intentional attacks on GNSS receivers might act at two different layers: directly on the receiver (*nonsignal attacks*) or at the GNSS signal-in-space (SIS) level (*signal attacks*).

The techniques based on a direct attack to the receiver belong to the first category (*nonsignal attacks*) and typically include tampering with the information within the receiver or the alteration of the position reported by the receiver to a service provider or a control center (i.e., man-in-the-middle attacks).

The second category (*signal attacks*) includes deliberate attacks on GNSS signals, which are conventionally categorized in three different forms [9, 10]:

- *Jamming.* Blocking reception of the GNSS signal by deliberately emitting electromagnetic radiation (i.e., radio-frequency interference) to disrupt user receivers by reducing the signal-to-noise level [11];
- *Meaconing.* Rebroadcasting of delayed GNSS signals without any distinction between SIS from different satellites [11, 12];
- *Spoofing.* Transmission of counterfeit GNSS-like signals, with the intent to produce a false position within the victim receiver without disrupting GNSS operations [12].

Note that jamming attacks, and other intentional interfering sources disrupting GNSS receiver operations, were discussed in Chapter 2. This chapter focuses on meaconing and spoofing, which represent a growing menace to current GNSS receivers and applications. These kinds of attacks are also known in the literature as *structured interferences*, since they are based on intentional transmission of delayed or counterfeit GNSS signals [13–18].

Assuming that a GNSS receiver is under attack, it receives a combination of both authentic signals (for the N_s satellites in view) and counterfeit signals (from a meaconer or a spoofer). Such a combined RF signal can be modeled at the input of the victim receiver as

$$y_{\mathrm{RF}}^{a}(t) = \sum_{l=0}^{N_s-1} s_{\mathrm{RF},l}(t) + n(t) + a(t) \tag{3.1}$$

where the a superscript in $y_{RF}^{a}(t)$ highlights the fact that the receiver is under attack, $s_{RF,l}(t)$ is the authentic signal received from the lth satellite in line of sight, $n(t)$ is the noise component (as defined in Chapter 1), and $a(t)$ represents an additional component received from the attacker. The $a(t)$ component includes the sum of all the counterfeit signals, which, basically, are amplified and delayed replicas of the authentic satellite signals. Thus, the component $a(t)$ can be modeled as

$$a(t) = \sum_{l=0}^{N_a-1} s'_{RF,l}(t) + n_a(t) \tag{3.2}$$

where N_a is the number of counterfeit signal replicas $s'_{RF,l}(t)$ that are generated by the attacker (usually $N_a \leq N_s$), and $n_a(t)$ represents a possible additional noise contribution due to the attacker. The $n_a(t)$ contribution increases the noise floor experienced by the victim receiver [i.e., $n(t) + n_a(t)$].

Each counterfeit signal $s'_{RF,l}(t)$ is usually an amplified and delayed replica of the signal $\hat{s}_{RF,l}(t)$ which represents an estimate of the authentic signal $s_{RF,l}(t)$ as obtained by the attacker from the lth satellite signal. The estimated signal $\hat{s}_{RF,l}(t)$ is potentially affected by inconsistencies with respect to $s_{RF,l}(t)$, due to possible demodulation/estimation errors occurring during the signal processing performed by the attacker. In addition, for the case of imperfect synchronization by the attacker (i.e., not locked on the authentic signals or showing residual errors in its Doppler frequency estimation), the counterfeit signal $s'_{RF,l}(t)$ possibly includes a residual modulation due to the frequency difference Δf_l and an initial phase difference $\Delta\theta_l$ between $s_{RF,l}(t)$ and $s'_{RF,l}(t)$.

For these reasons, a general model of $s'_{RF,l}(t)$ can be written as

$$s'_{RF,l}(t) = A_l(t)\hat{s}_{RF,l}\left(t - \tau_l(t)\right)\cos\left(2\pi\Delta f_l t + \Delta\theta_l\right) \tag{3.3}$$

where $A_l(t)$ represents the ratio between the amplitude of the counterfeit replica and the authentic signal, and $\tau_l(t)$ is the relative delay between them. Note that, depending on the type of attack, both $A_l(t)$ and $\tau_l(t)$ can vary over time or can be constant values.

After this brief introduction, this chapter describes different types of meaconing and spoofing attacks and, for each category of attack, a different analytical model for the component $a(t)$ is provided, discussing the related parameters. Special attention is also devoted to possible hybrid/combined techniques. The main characteristics of each type of attack are outlined in an attempt to answer the following questions:

- What does the attack consists of?
- How is the attack impacting on the target receiver?
- What is required in order to implement it?
- Are there any limits or drawbacks at the attacker side?

Possible countermeasures able to detect and or mitigate each specific attack will be discussed in Chapter 8.

3.2 Meaconing

Meaconing is defined as the reception, delay, and rebroadcast (or the recording and playback) of an entire block of the radio-frequency (RF) spectrum containing an ensemble of GNSS signals [18, 19]. As summarized in Table 3.1, the GNSS signals received from the satellites in view are not typically separated during the reception phase. Furthermore, the meaconing attack introduces a relative delay such that the meaconed signal arrives at the target receiver with a positive delay with respect to the authentic signal.

Note that the number of counterfeit signal replicas in a meaconing attack corresponds to the number of satellites in view ($N_a = N_s$). In addition, the meaconer generates counterfeit signal replicas $s'_{\mathrm{RF},l}(t)$ that correspond exactly with the authentic signals $s_{\mathrm{RF},l}(t)$ (i.e., $\hat{s}_{\mathrm{RF},l}(t) = s_{\mathrm{RF},l}(t)$), apart from a constant amplification factor ($A_l(t) = A_m > 1$) and a constant delay [$\tau_l(t) = \tau_m > 0$]. It also introduces an additional noise contribution ($n_a(t) = n_m(t)$) due to its hardware components (antenna, LNA, and transmitting front-end, as mentioned in Table 3.1). This contribution is not negligible and increases the total noise power seen by the victim receiver (i.e., $n(t) + n_m(t)$). The meaconer does not introduce a residual modulation on the counterfeit signals (i.e., $\Delta f_l = 0$ and $\Delta\theta_l = 0$). At this point, (3.2) can be rewritten as

$$a(t) = \sum_{l=0}^{N_s-1} A_m s_{\mathrm{RF},l}\left(t - \tau_m\right) + n_m(t) \tag{3.4}$$

where the subscripts m highlight the specific parameters of the meaconing attack. Note that, in the case of digital signal processing at the meaconer, the delay τ_m can also be expressed as a positive number of samples.

Table 3.1
Main Characteristics of a Meaconing Attack

Description	Reception, delay, and rebroadcast of radionavigation signals.
Hardware requirements	GNSS antenna + low noise amplifier (LNA) + RF transmitting front-end. Alternative: recording and playback system.
Impact on the target receiver	It cannot arbitrarily manipulate the PVT of target receivers. Rather, target receivers will be confused and will display the position and velocity of the meaconer and a time in arrears of true time.
Context requirements	It has to be located in proximity to the target receiver, or the LNA gain must be calculated and adapted in order to ensure a proper power level, depending on the distance between the spoofer and the target receiver.
Limits of implementation	Easy to implement; it requires few RF components, without specific software developments.
Difficulty of detection	Possibly detectable, if the introduced delay is not consistent with the receiver clock drift (also depending on the last valid PVT solution).

An example of a meaconing attack can be found in [6] and [20] where the authors utilize a simple repeater for live testing.

3.3 Spoofing

As shown in Figure 3.1, spoofing attacks are often classified in the GNSS literature as *simplistic*, *intermediate* (or shadowed), and *sophisticated* [16, 17]. Other references use different nomenclatures and categorizations of the attacks, depending on the specific features and implementation details of each attack. For example, a key feature is the possible *synchronization* of the counterfeit signals with respect to the authentic ones (i.e., synchronized or unsynchronized spoofers). However, the three categories of simplistic, intermediate, and sophisticated spoofing provide a general classification of the attacks that depends on the implementation complexity. Such a high level of abstraction can be further enriched by adding contingent specifications for each type of threat, including the technical details and also mentioning possible variants for each category. Thus, these three categories are adopted and discussed in following subsections. In addition, for the sake of comprehensiveness, modified/hybrid approaches that fall outside of these conventional categories are also considered in Section 3.4.

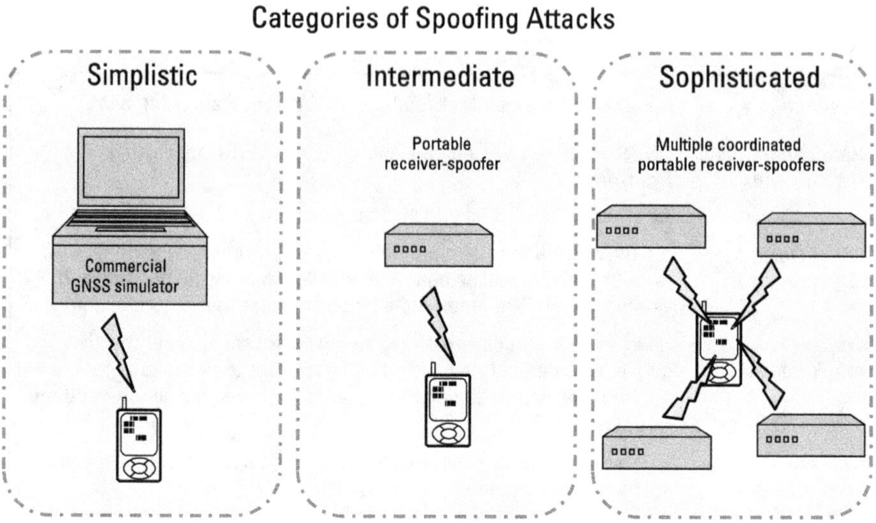

Figure 3.1 Illustration of simplistic, intermediate, and sophisticated spoofing attacks (inspired by [16]).

3.3.1 Simplistic Attack

In a simplistic spoofing attack, a GNSS signal simulator is typically concatenated with an RF transmitting front-end and is employed to mimic authentic signals (see Table 3.2). A simplistic spoofer can generate GNSS signals, but it is generally not able to make them consistent and time synchronized with the real signals. It is considered the simplest spoofer but, because the generated signals are not synchronized to the real ones and have higher power, this type of spoofing can be tackled with simple countermeasures.

In case of a simplistic attack, the counterfeit signals can be modeled as in (3.2) and (3.3). Note that the counterfeit signals $s'_{RF,l}(t)$ can be inconsistent with respect to the authentic signals $s_{RF,l}(t)$. In fact, they can be affected by rough discrepancies in the navigation messages (e.g., old ephemeris data) and/or they can represent satellites not visible at the location of the victim receiver. In addition, due to the missing synchronization, they are typically affected by a nonnegligible residual modulation effect (i.e., $\Delta f_l \neq 0$ and $\Delta\tau_l \neq 0$).

As an example, a successful simplistic attack has been carried out by the authors and the results are reported in Figure 3.2. In this case, an experimental test was performed using a consumer-grade receiver and a commercial hardware GNSS signal simulator. The test started from an initial condition with the

Table 3.2
Main Characteristics of a Simplistic Spoofing Attack

Description	A GPS signal simulator broadcasts high-powered counterfeit GPS signals toward a victim receiver.
Hardware requirements	A GPS signal simulator, combined with a power amplifier and an RF transmitting front-end.
Impact on the target receiver	The spoofing signals look like noise for a receiver operating in the tracking mode (the spoofing signals are not essentially synchronized to the real signals). It may cause the victim receiver to lose lock and undergo to a partial or complete reacquisition.
Context requirements	Broadcasting antenna situated close to the target receiver antenna or spoofer directly connected to the antenna cable of the victim receiver, in case of complicit spoofing.
Limits of implementation	Easy to implement; it requires commercial components only, without specific software developments. Cost and size: most signal simulators are expensive, heavy, and cumbersome.
Difficulty of detection	Easy to detect, due to the difficulty of synchronizing a simulator's output with the actual GNSS signals in its vicinity, leading to possible jumps in its PVT solution.

receiver tracking live GPS signals from a rooftop antenna (located in Turin, Italy; latitude 45.065274353° N, longitude 7.6589692° E, height 312 m). The receiver under test was forced to unlock from the real signals by unplugging its antenna cable. This cable was then directly plugged in to the signal simulator, which was configured in order to emulate the received signals in a different location (Rome; latitude 41.893056° N, longitude 12.482778° E, height 21 m, at a distance of about 525 km from Turin). In addition, the timescale of the simulator was misaligned with respect to the real GPS time (the difference was 1 min). This case is representative of a complicit spoofing scenario, where the attacker has complete access to the cable or the antenna location.

Focusing on the outputs of the receiver reported in Figure 3.2, notice the discontinuities in the estimated position and in the timescale during the test. In detail, the position reported by the receiver jumps from the real location (Turin) to the simulated position (Rome), as shown in Figure 3.2(a). In addition, due to the misalignment between real and simulated timescales (1 min in this example), the time information logged from the receiver shows a discontinuity, as highlighted in Figure 3.2(b). The missing time interval can also be noticed on the horizontal axes of both Figures 3.2(c) and (d), where

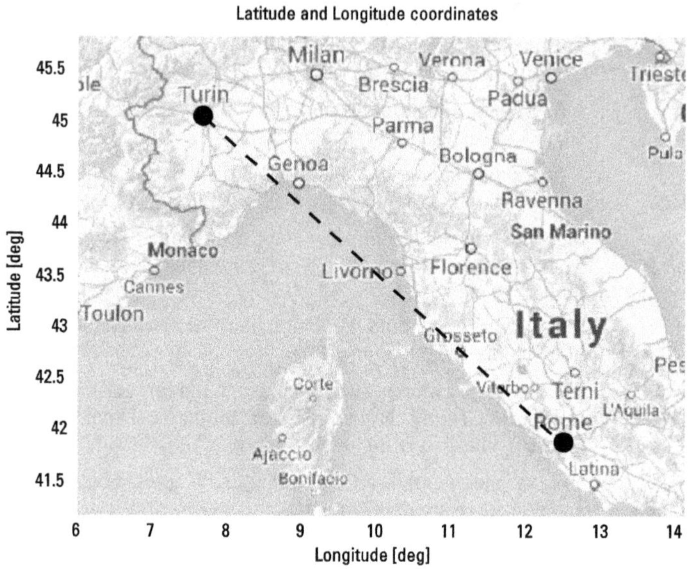

(a)

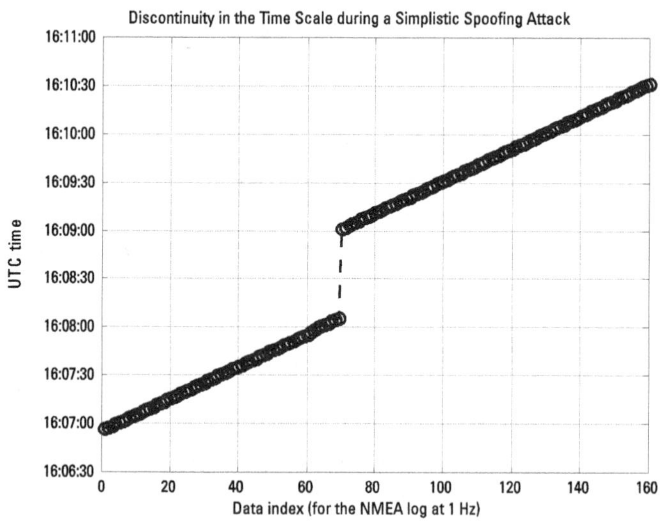

(b)

Figure 3.2 Results of a simplistic spoofing attack: (a) real and counterfeit position of the target receiver and discontinuities in the (b) timescale.

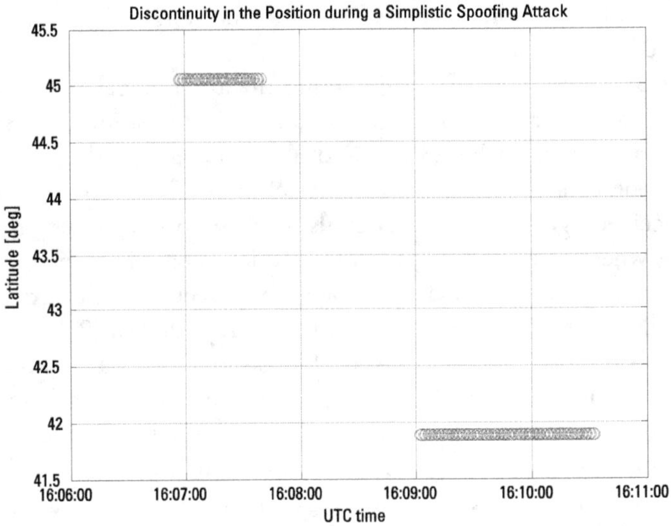

(c)

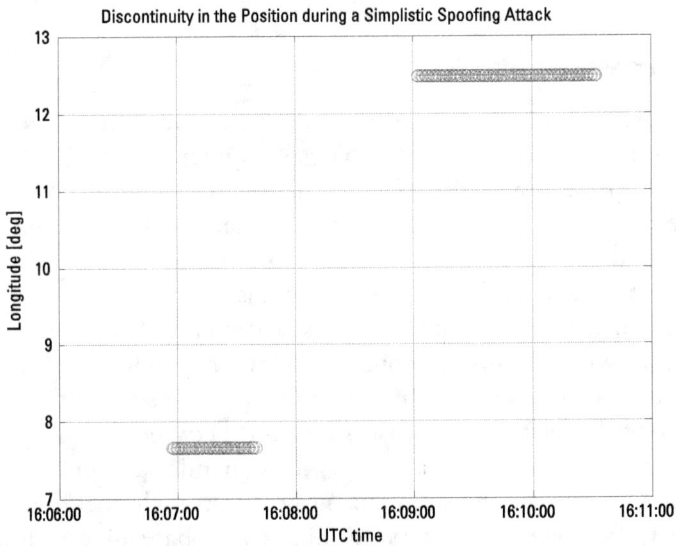

(d)

Figure 3.2 *(Continued)* (c) estimated latitude, and (d) longitude.

the discontinuities on the estimated latitude and longitude coordinates are also reported.

Note also that, just after unplugging the antenna cable and plugging in the simulator, the receiver reported several invalid positions. However, the discontinuities shown in Figure 3.2 and, more in general, the behavior of a receiver in the presence of a spoofing attack depends on the internal logic and the receiver algorithms. More details on these topics will be provided in Chapter 8, where possible antispoofing techniques based on the monitoring of the measurements and the outputs of a GNSS receiver will be discussed.

Another example of a simplistic attack is reported in [21]. Researchers at Argonne National Laboratories were able to spoof two mass market receivers (commonly used for hiking and automotive) into reporting false position information. In [21], they describe how it is necessary to block the victim receiver signal before starting the actual spoofing attack. They did so by obscuring the antenna. Alternatively, a GPS jammer could be used in order to force the victim receiver to unlock from the true signal and then to lock on to the (stronger) counterfeit signal.

3.3.2 Intermediate Attack

This type of spoofing attack is more sophisticated and more sinister than the simplistic one. As summarized in Table 3.3, an intermediate spoofer consists of a device capable of receiving GNSS signals and generating counterfeit signals. The spoofer extracts time, position, and satellite information from the received signals, and then leverages the synchronization of the local codes and carriers to generate plausible counterfeit signals.

In an intermediate spoofing attack, the counterfeit signals are code-phase aligned with the authentic ones. In detail, they should be synchronized at least within a half code chip in order to carry out a successful attack. The knowledge of the relative positions (3D vector) between the spoofer and the target antennas is mandatory, as well as the dynamics. As illustrated in Figure 3.3, the receiver-spoofer simultaneously attacks each tracking channel of the target GPS receiver by first performing code-phase alignment and then signal lift-off [16, 17].

Recalling (3.2) and (3.3), the behavior shown in Figure 3.3 can be obtained by means of a proper variation in time of the relative amplitude ratio $A_l(t)$ and the relative delay $\tau_l(t)$ between the counterfeit and the authentic signal from each lth satellite in view. It must also be noticed that an intermediate spoofer is usually synchronized with respect to the authentic signals, thus the Δf_l parameter in (3.3) is negligible (i.e., $\Delta f_l = 0$).

Table 3.3
Main Characteristics of an Intermediate Spoofing Attack

Description	This spoofing device first synchronizes to live GNSS signals, and then generates the spoofing signal knowing the 3D pointing vector of its transmit antenna toward the target receiver antenna.
Hardware requirements	A custom device properly designed for spoofing purposes or, as an alternative, a modified GNSS receiver combined with an RF transmitting front-end.
Impact on the target receiver	Each channel of the target receiver is brought under control. The counterfeit correlation peak is aligned with the peak corresponding to the genuine signal. The power of the counterfeit signal is then gradually increased ("shadowed" spoofing). Eventually, the counterfeit signal gains control of the DLL tracking points that flank the correlation peak (see Figure 3.3).
Context requirements	It requires accurate knowledge of the target receiver's antenna position and velocity (dynamics). Self-spoofing (known as limpet spoofing) can be carried out easily. In fact, the spoofer can be made small enough to be placed inconspicuously near the target receiver's antenna.
Limits of implementation	Fairly sophisticated software to be implemented in the spoofer. Low-cost hardware.
Difficulty of detection	Difficult to detect and mitigate; only complex countermeasures (e.g., angle-of-arrival defense) are effective against an intermediate attack.

A possible implementation of a portable civilian GPS spoofer is described in [16]. The authors present the design of an intermediate spoofer, implemented as a modified software-defined receiver combined with an RF transmitting front-end. This receiver-spoofer is able to perform an attack that would defeat most known user-equipment-based spoofing countermeasures.

In addition to an architecture based on a modified GNSS receiver as in [16], intermediate attacks have been emulated by means of testbeds [22–24]; these setups allow researchers to experimentally test possible attacks and to validate the proposed countermeasures in repeatable scenarios.

It must be pointed out that the prediction of the navigation data bits is needed for decreasing the effectiveness of possible countermeasures against intermediate attacks, avoiding possible inconsistencies between $\hat{s}_{RF,l}(t)$ and $s_{RF,l}(t)$ in (3.2) and (3.3). A specific option for the implementation of an attack exploiting on-the-fly bit estimation is the security code estimation and replay (SCER) attack, which will be discussed in Section 3.4.

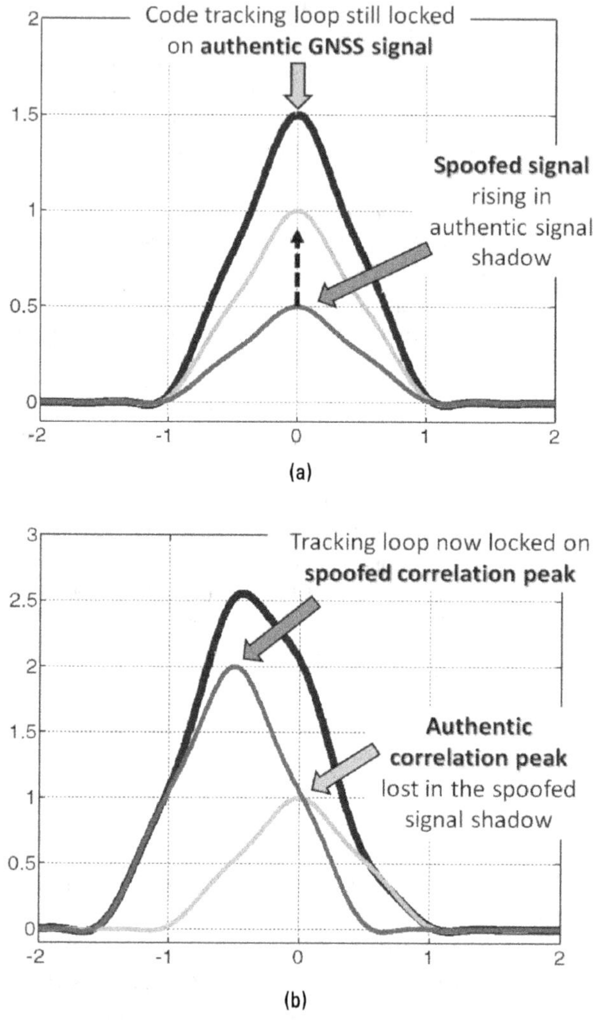

Figure 3.3 Illustration of an intermediate (shadowed) spoofing attack on a single channel of a GNSS receiver (inspired by [16]). The counterfeit signal is aligned to the authentic signal and its power is increased (a) until the spoofer gains control on the tracking loop (b).

3.3.3 Sophisticated Spoofers

Multiple, coordinated and synchronized receivers might become the most sophisticated spoofing device [17]. As summarized in Table 3.4, these coordinated receivers are able to generate and transmit counterfeit signals as in the case of intermediate spoofing. In addition, they have sub-centimeter-level 3D position information about their antennas' phase centers and the target antenna's phase center and they can readily defeat complex countermeasures (such as the angle-of-arrival defense) by relying on the constructive properties of their RF signals. Furthermore, these spoofers can suppress authentic signals at the target receiver's antenna. Thus, only a cryptographic defense (such as signal authentication) or the use of hybrid solutions (leveraging data from other sensors) represents strong countermeasures against this type of spoofer, as will be discussed in Chapter 8.

Recalling (3.2) and (3.3), note that a sophisticated spoofing attack is characterized by accurate synchronization between the counterfeit and authentic signals, leading to negligible frequency and phase differences (i.e., $\Delta f_l = 0$ and $\Delta\theta_l = 0$).

Unsurprisingly, development and fielding of such sophisticated devices is challenging, because of the complexity associated with the development of this kind of spoofer and the need for sub-centimeter-level knowledge of the target receiver's antenna. These facts make the achievement of a coordinated attack with sophisticated spoofers even harder and then its likelihood is relatively low [17]. In fact, authors of [8] specify that no open literature has reported development of sophisticated attacks.

Table 3.4
Main Characteristics of a Sophisticated Spoofing Attack

Description	A network of coordinated intermediate spoofers replicates not only the content and mutual alignment of visible signals, but also their spatial distributions.
Hardware requirements	Multiple phase-locked portable receiver-spoofers (i.e., intermediate spoofers; see Table 3.3).
Impact on the target receiver	Similar to the intermediate spoofing attack (see Table 3.3). Most effective spoofing category.
Context requirements	Sub-centimeter-level knowledge of the position and velocity of the target receiver antenna phase center.
Limits of implementation	Most complex spoofing category. The effectiveness region is much more limited.
Difficulty of detection	Able to fool even multiple-antenna (angle-of-arrival) spoofing defenses. It may be impossible to detect with GNSS-only-based spoofing defenses.

3.4 Hybrid/Combined Spoofing Techniques

The conventional classification of the spoofing attacks presented in the previous subsections (i.e., simplistic, intermediate and sophisticated) is recommended in order to be able to compare different techniques. However, it may be too restrictive in some practical cases; in fact, modified versions of the conventional attacks and hybrid/combined solutions are already being mentioned in literature. Thus, in order to provide as comprehensive an analysis as possible, hybrid/combined techniques are outlined next.

3.4.1 Relaying Attack

As mentioned in [15], the relaying attack (or wormhole attack) represents a modified version of the classical meaconing attack (previously discussed in Section 3.2).

As illustrated in Figure 3.4, the relaying attack differs from a conventional meaconing because the receiving antenna and the RF transmitting front-end of the meaconer (including a low-noise amplifier and a transmitting antenna) are located far away in distant places. In this case their distance makes infeasible a conventional meaconing attack (i.e., a direct reception and rebroadcast of the GNSS signal), because it will require excessive transmission power (i.e., LNA gain). For this reason, a real-time radio link is used for transmitting the GNSS signals received from the remote antenna to the target receiver. Two modems can be used in order to perform the analog-to-digital conversion (ADC) and then the digital-to-analog conversion (DAC) of the signal, as in the lower part of Figure 3.4.

The counterfeit signal for a relaying attack can be modeled in a similar way as previously done in (3.4) for the conventional meaconing:

$$a(t) = \sum_{l=0}^{N_a-1} A_r \hat{s}_{\mathrm{RF},l}\left(t-\tau_r\right) + n_r(t) \tag{3.5}$$

where the subscripts r highlight the specific parameters of the relaying attack, which are the amplification factor A_r, the delay τ_r, and the noise contribution $n_r(t)$. Note that the counterfeit signals $\hat{s}_{\mathrm{RF},l}(t)$ can be inconsistent with respect to the authentic signals $s_{\mathrm{RF},l}(t)$, due to the large distance between the remote antenna and the victim receiver and then to the different visible satellites ($N_a \neq N_s$). Another important difference is due to the fact that, in the case of a relaying attack, the noise component $n_r(t)$ added by the attacker also takes into account the noise effects of the radio link (including the two modems).

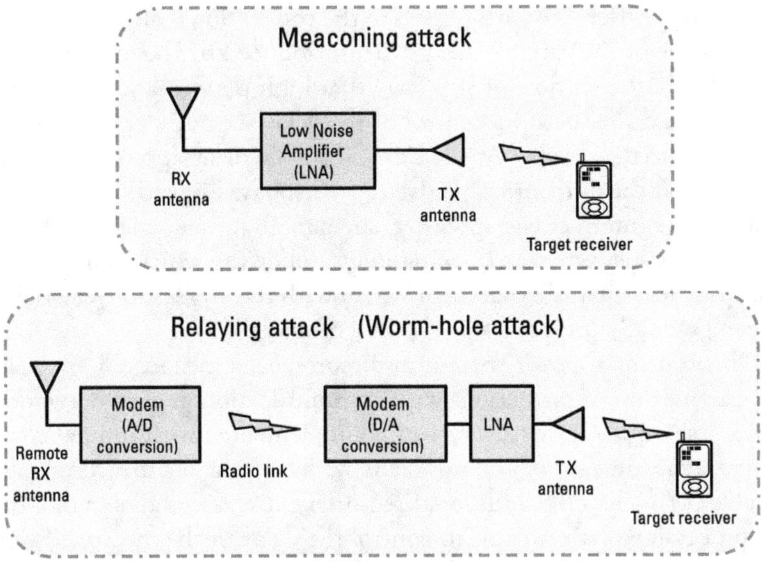

Figure 3.4 Differences between a conventional meaconing and a relaying attack.

Such a technique might be quite hard to detect with conventional countermeasures based on GNSS signals only. To detect the attack, it might be necessary to cross-check the validity of the PVT solution with other sensors (e.g., a highly stable clock, for detecting possible inconsistencies in the timescale). However, such an attack may be logistically complex, especially if the target receiver is not static; in fact, arrangements may have to be made to move the remote antenna in a plausible way with respect to the target antenna, without being noticed.

3.4.2 Meaconing with Variable Delay

Another modified version of the conventional meaconing attack is mentioned in [13]. Instead of having a constant delay as in (3.4) or (3.5), the idea is to control the delay inserted by the meaconer. This allows for the fooling of possible implemented countermeasures based, for example, on the monitoring of the clock drift. At this point, this type of attack can be modeled as

$$a(t) = \sum_{l=0}^{N_s-1} A_m s_{\mathrm{RF},l}\left(t - \tau_m(t)\right) + n_m(t) \tag{3.6}$$

where the only difference with respect to the conventional meaconing in (3.4) is the presence of the variable delay $\tau_m(t)$ introduced by the meaconer.

In detail, the authors of [13] state that high-performance digital signal processing hardware permits control of the delay (in number of digital samples) to ever smaller values. At the limit, if such a delay approaches zero, the meaconed and the authentic signals are code-phase-aligned. They conclude that, as for the intermediate spoofing, such an alignment enables a seamless liftoff of the target receiver's tracking loops, following which a meaconer can increase the delay at a rate that is consistent with the target receiver clock drift and then gradually impose a significant timing delay.

This type of attack is specific and represents a menace for applications based on the time information (e.g., communication/energy network synchronization). In fact, this attack can be implemented with limited cost and resources, requiring similar components with respect to a meaconing attack (see Table 3.1). The only additional requirement is the availably of a digital signal processing platform able to control the delay of the meaconed signal.

In addition, a meaconer with a variable delay can be combined with other techniques in a cascade for more sophisticated attacks based on multiple steps. For example, it is possible to use it as a first step in order to furtively insert a delay on all the received GNSS signals. Once such a delay is sufficient for performing a reliable estimation of the received navigation data bits, a second step of the attack can begin, including harmful arbitrary manipulations of the received signal (e.g., by means of a SCER attack, as discussed in following section).

3.4.3 Security Code Estimation and Replay Attack

An option for the implementation of an intermediate spoofing attack is presented in [13] and [18]. It is called a security code estimation and replay (SCER) attack and can be used against GNSS signals featuring cryptographic defenses (i.e., unpredictable navigation message bits or security code chips). The idea of the SCER attack is to estimate (and not only predict) on the fly the values of each of the encrypted and unpredictable code chips (or navigation message bits) that are present in the received signal.

Recalling previous expressions in (3.2) and (3.3), the main difference between an SCER attack and conventional intermediate spoofers is that the security code chips (or navigation message bits) in $\hat{s}_{\mathrm{RF},l}(t)$ are not predicted, but are estimated on the fly. This means that, in order to be able to generate a plausible counterfeit signal $s'_{\mathrm{RF},l}(t)$, the spoofer has to directly observe the spreading code sequence and the navigation data bits from the received signal

$s_{RF,l}(t)$. This estimation process forces the spoofer to introduce a positive observation delay $\tau_{obs,l}$ between the received signal and the counterfeit replica. This constant delay ($\tau_{obs,l} > 0$) is needed in order to ensure a reliable estimation of the received signal, avoiding inconsistencies in $s'_{RF,l}(t)$. An additional control delay $\tau_{ctrl,l}(t)$ is also needed by the spoofer in order to be able to control the relative delays between the signals from different satellites.

For these reasons, the overall delay $\tau_l(t)$ between the counterfeit replica and the authentic signal from the *l*th satellite can be written as

$$\tau_l(t) = \tau_{obs,l} + \tau_{ctrl,l}(t) \tag{3.7}$$

where $\tau_{ctrl,l}(t)$ can be independently modified by the spoofer in order to manipulate the PVT solution of the target receiver.

As discussed in [13] and [18], where the mathematical details of the method are also provided, such estimation and control delay between the spoofing and the authentic signal is a key aspect of the SCER attack and it allows overcoming possible cryptographic defenses.

However, the estimation of the security code chips increases the complexity and associated costs of SCER with respect to a conventional intermediate spoofing attack.

3.4.4 Meaconing or Spoofing Plus High-Gain Antennas

A more sophisticated spoofing attack effective against cryptographically protected signals (see Chapter 8) is based on the use of high-gain GNSS antennas, as described in [15]. In this case, at least four directional antennas are pointed at four different satellites and employed to separately demodulate their signals $s_{RF,l}(t)$. If the antenna gain is sufficient, data bits and spreading codes rise above the noise floor and can be directly observed without the despreading process.

Recalling (3.2) and (3.3), thanks to the high-gain antennas that lift the observed signal above the noise floor, each estimated signal $\hat{s}_{RF,l}(t)$ for each *l*th satellite is not affected by inconsistencies and estimation errors. Thus, the spoofing device can regenerate each satellite signal component, producing a plausible counterfeit signal $s'_{RF,l}(t)$. As in other types of attacks, selective delays $\tau_l(t)$ can be inserted in order to manipulate the target receiver's PVT solution.

However, depending on the gain of the directional antennas and on the signal characteristics, the achievable signal-to-noise ratio (SNR) might not be sufficient for directly separating and decoding the signal components in a reliable way [15]. In such a situation, instead of a spoofing attack, a modified meaconing attack is still feasible by selectively delaying the raw signals from each antenna and then mixing these channels together before rebroadcasting the combined signal. The composite meaconing signal can be modeled as

$$a(t) = \sum_{k=0}^{N_a-1} A_{m,k}\left[s_{\mathrm{RF},k}\left(t-\tau_k(t)\right)+n_{m,k}\left(t-\tau_k(t)\right)\right] \tag{3.8}$$

where N_a is the number of high-gain antennas pointed at different satellites, corresponding to different channels combined by the meaconer, and $A_{m,k}$, $\tau_k(t)$, and $n_{m,k}(t)$ are the amplification factor, the delay, and the noise contribution corresponding to the kth channel, respectively.

The main drawback of such modified meaconing with respect to previous multiple-antenna spoofing is that all of the noise components $n_{m,k}(t)$ related to the raw signals from each kth antenna are summed in $a(t)$, increasing the total noise contribution seen by the victim receiver (i.e., $n(t) + \sum_{k=0}^{N_a-1} n_{m,k}(t)$). In addition, for each kth channel, the satellite signals not completely suppressed by the antenna gain (i.e., near the directional antenna beam) will sum up to the desired (counterfeit) satellite signal, providing potential clues for antispoofing countermeasures.

From a theoretical point of view, the use of high-gain antennas helps to overcome cryptographic defenses, and therefore both spoofing and meaconing attacks can be accomplished. However, a practical implementation of these attacks is not straightforward, basically due to the high cost and the logistically complex issues related to the setup of multiple high-gain antennas for separately receiving the signals from each satellite in view. These problems reduce the applicability and the feasibility of such an attack, especially in the case of a large number of satellites in view and if the target receiver is not static.

3.5 Conclusions

Possible menaces and vulnerabilities at the GNSS SIS level have been considered in this chapter. In particular, this chapter focused on possible attacks that aim at fooling a GNSS receiver, forcing it to provide erroneous solutions.

Different types of meaconing and spoofing have been categorized and discussed. These attacks are also known in the literature as structured interferences, since they are based on intentional transmission of delayed or counterfeit GNSS signals. The conventional classification of simplistic, intermediate, and sophisticated attacks was adopted in the first part of the chapter in order to have a common terminology. However, hybrid attacks, which do not fit directly into conventional categories, are recalled for the sake of completeness. They include modifications and/or combinations of classical meaconing and spoofing approaches. These hybrid techniques result in the most dangerous type of attacks. In fact, multiple techniques can be combined in a cascade, leading to sophisticated attacks based on multiple steps. Conventional detection and mitigation solutions can be ineffective in the presence of these hybrid attacks, thus proper countermeasures are needed, as will be discussed in Chapter 8.

References

[1] Linty, N., and P. Crosta, "Code and Frequency Estimation in Galileo Mass Market Receivers," *Proc. 2013 Int. Conf. on Localization and GNSS (ICL-GNSS)*, June 25–27, 2013, pp. 1–6. doi:10.1109/ICL-GNSS.2013.6577262

[2] Heng, L., D. B. Work, and G. X. Gao, "Cooperative GNSS Authentication. Reliability from Unreliable Peers," *Inside GNSS*, Vol. 8, No. 5, September/October 2013, pp. 70–75.

[3] Global Positioning Systems Directorate Systems Engineering & Integration, *Navstar GPS Space Segment/Navigation User Interface*, Interface Specification IS-GPS-200H, September 24, 2013.

[4] European Commission, *European GNSS (Galileo) Open Service Signal in Space Interface Control Document*, OS SIS ICD, Issue 1.1, September 2010.

[5] Russian Institute of Space Device Engineering, *Global Navigation Satellite System GLONASS Interface Control Document*, Navigational Radiosignal in Bands L1, L2, Edition 5.1, 2008.

[6] Akos, D. M., "Who's Afraid of the Spoofer? GPS/GNSS Spoofing Detection via Automatic Gain Control (AGC)," *J. of the Institute of Navigation*, Vol. 59, No. 4, Winter 2012, pp. 281–290.

[7] Troglia Gamba, M., B. Motella, and M. Pini, "Statistical Test Applied to Detect Distortions of GNSS Signals," *Proc. 2013 Int. Conf. on Localization and GNSS (ICL-GNSS)*, June 25–27, 2013, pp. 1–6. doi:10.1109/ICL-GNSS.2013.6577267.

[8] Wesson, K., D. Shepard, and T. Humphreys, "Straight Talk on Anti-Spoofing. Securing the Future of PNT," *GPS World*, Vol. 13, No. 1, January 2012, pp. 32–63.

[9] Royal Academy of Engineering, *Global Navigation Space Systems: Reliance and Vulnerabilities*, March 2011.

[10] Wildemeersch, M., et al., "Impact Study of Unintentional Interference on GNSS Receivers," Technical Report of the EC Joint Research Center, Security Technology Assessment Unit, EUR 24742 EN, 2010.

[11] De Castro, H. V., G. van der Maarel, and E. Safipour, "The Possibility and Added-value of Authentication in future Galileo Open Signal," *Proc. 23rd Int. Technical Meeting of the Satellite Division of the Institute of Navigation*, Portland, OR, September 2010.

[12] Kaplan, D. E., and C. J. Hegarty, *Understanding GPS: Principles and Applications*, 2nd ed., Norwood, MA: Artech House, 2006.

[13] Wesson, K., M. Rothlisberger, and T. Humphreys, "Practical Cryptographic Civil GPS Signal Authentication," *J. of the Institute of Navigation*, Vol. 59, No. 3, Fall 2012, pp. 177–193.

[14] Lo, S., et al., "Signal Authentication: A Secure Civil GNSS for Today," *Inside GNSS*, Vol. 4, No. 5, September/October 2009, pp. 30–39.

[15] Kuhn, M. G., "An Asymmetric Security Mechanism for Navigation Signals," *Information Hiding, Lecture Notes in Computer Science*, Vol. 3200, Berlin, Germany: Springer, 2005, pp. 239–252.

[16] Humphreys, T., et al., "Assessing the Spoofing Threat: Development of a Portable GPS Civilian Spoofer," *Proc. 21st Int. Technical Meeting of the Satellite Division of the Institute of Navigation* (ION GNSS 2008), Savannah, GA, September 2008, pp. 2314–2325.

[17] Ledvina, B. M., et al., "An In-Line Anti-Spoofing Device for Legacy Civil GPS Receivers," *Proc. 2010 Int. Technical Meeting of the Institute of Navigation*, San Diego, CA, January 2010, pp. 698–712.

[18] Wesson, K., M. Rothlisberger, and T. Humphreys, "A Proposed Navigation Message Authentication Implementation for Civil GPS Anti-Spoofing," *Proc. 24th Int. Technical Meeting of the Satellite Division of the Institute of Navigation* (ION GNSS 2011), Portland, OR, September 2011, pp. 3129–3140.

[19] John A. Volpe National Transportation Systems Center, *Vulnerability Assessment of the Transportation Infrastructure Relying on the Global Positioning System*, Final Report, August 2001.

[20] Akos, D. M., "GNSS RFI/Spoofing: Detection, Localization, & Mitigation," paper presented at Stanford's 2012 PNT Challenges and Opportunities Symposium, November 2012.

[21] Warner, J. S., and R. G. Johnston, "A Simple Demonstration That the Global Positioning System (GPS) Is Vulnerable to Spoofing," *Journal of Security Administration*, Vol. 25, 2002, pp. 19–28.

[22] Humphreys, T., et al., "The Texas Spoofing Test Battery: Toward a Standard for Evaluating GPS Signal Authentication Techniques," *Proc. 25th Int. Technical Meeting of the Satellite Division of the Institute of Navigation* (ION GNSS 2012), Nashville, TN, September 2012, pp. 3569–3583.

[23] Pozzobon O., et al., "Status of Signal Authentication Activities within the GNSS Authentication and User Protection System Simulator (GAUPSS) Project," *Proc. 25th Int. Technical Meeting of the Satellite Division of the Institute of Navigation* (ION GNSS 2012), Nashville, TN, September 2012, pp. 2894–2900.

[24] Humphreys, T., et al., "A Testbed for Developing and Evaluating GNSS Signal Authentication Techniques," *Proc. Int. Symp. on Certification of GNSS Systems & Services* (CERGAL), Dresden, Germany, 2014, pp. 1–15.

4

Analytical Assessment of Interference on GNSS Signals

Fabio Dovis, Luciano Musumeci, Davide Margaria, and Beatrice Motella

4.1 Introduction

As described in Chapter 3, radio-frequency interference (RFI) is a category that includes several different signals with various characteristics that can affect satellite navigation receivers in very different ways. For such a reason it is important to have analytical tools that allow for, particularly during the design phase, the assessment of expected performance with respect to the interfering signal features and the receiver setup.

Measurement of the carrier–to–noise density ratio (C/N_0) is fundamental to determining the performance quality of a GNSS receiver, because it is commonly accepted as the parameter versus which the different metrics are represented (e.g., carrier and code tracking jitter). When a signal is affected by the presence of interference, a loss in performance is experienced (see Chapter 2) and it is useful to define an effective C/N_0, taking into account equivalence with the actual C/N_0 that would lead to the same loss in performance.

However, the effect of the presence of interference cannot be modeled only as an increase of "equivalent noise." In fact, depending on the nature of the interfering signals, and also taking into account the receiver processing, the impact on the pseudorange estimation may cause an increased variance as well as a bias of the measurements, thus leading to a loss in accuracy in the position estimation.

In this chapter we introduce a classical method based on the spectral separation coefficient (SSC), which is useful for estimating the loss in the C/N_0 estimated by the receiver, and the most recent technique of the interference error envelope, which allows for a prediction of the measurement bias.

4.2 Theoretical Model of the C/N_0 Loss in the Presence of Interference

A detailed derivation of the effective C/N_0 in the presence of narrowband interference is provided in [1] and [2]. The proposed model assumes a complex-valued baseband signal $s(t)$ with an unknown received delay τ and phase θ; the frequency is known within a fraction of the reciprocal of the coherent integration time. Naming $w(t)$ the interference plus noise component, and assuming $s_1(t)$ to be the useful GNSS signal filtered by the transmitter transfer function $H_T(f)$ and the receiver transfer function $H_R(f)$ (which also filters the noise component), the received signal becomes

$$x(t) = e^{j\theta} s_1(t-\tau) + w(t) \tag{4.1}$$

Another important assumption is that no automatic gain control (AGC) counter is triggered and therefore the gain can be considered constant. The received signal has a power spectrum $CG_s(f)$, where $G_s(f)$ is the power spectral density (PSD) normalized to unit area over infinite limits and C is the received signal power. The thermal noise has power spectral density N_0 and the interference has power spectral density $C_I G_I(f)$ with $\int_{-\beta_r/2}^{\beta_r/2} G_I(f) = 1$, where β_r is the front-end bandwidth. Furthermore, the gain of the receiver front-end is assumed to be known.

In [1] the effective C/N_0 is estimated starting from the postcorrelation Signal-to-noise ratio (SNR). Because in both coherent and noncoherent architectures the same formulation of the effective C/N_0 is achieved, the most significant steps of such a mathematical derivation are reported hereafter for the coherent signal-to-noise SNR case.

The coherent SNR is obtained as the ratio between the mean and the variance of a coherent test statistic $\lambda(\tau,\theta)$, which is the real part of the prompt correlation between the received data and the locally generated replica *s*(*t*). Thus, the test statistic becomes

$$\lambda(\tau,\theta) = \Re\left\{\frac{1}{T}\int_{-T/2}^{T/2} e^{i\theta} s_1(t)s^*(t-\tau)\,dt + \frac{1}{T}\int_{-T/2}^{T/2} w(t)s^*(t-\tau)\,dt\right\} \quad (4.2)$$

where the first integral is the signal component, while the second integral refers to the noise component in the definition of the prompt correlators and T denotes the integration time in which the received data are correlated with the local replica. Since $w(t)$ contains the noise plus the interference components, it is possible to define a coherent signal-to-noise plus interference ratio (SNIR) as

$$\rho_c = \frac{\left|E\{\lambda(\tau,\theta)\}\right|^2}{\mathrm{var}\{\lambda(\tau,\theta)\}} = \frac{2TC\left[\Re\left\{e^{i\theta}\int_{-\infty}^{\infty} G_S(f)H_T(f)H_R(f)e^{i2\pi f\tau}\right\}\right]^2}{\int_{-\infty}^{\infty} G_w(f)G_S(f)\,df} \quad (4.3)$$

The output SNIR achieves a maximum at $\tau = 0$. Thus, splitting the noise and the interference components as $G_w(F) = \left|H_R(f)\right|^2\left[N_0 + C_I G_I(f)\right]$, the final expression of the SNIR becomes

$$\rho_c = \frac{2T\dfrac{C}{N_0}\left[\int_{-\infty}^{\infty}\left|H_R(f)\right|^2 G_S(f)\,df\right]^2}{\int_{-\infty}^{\infty}\left|H_R(f)\right|^2 G_S(f)\,df + \dfrac{C_I}{N_0}\int_{-\infty}^{\infty}\left|H_R(f)\right|^2 G_I(f)G_S(f)\,df} \quad (4.4)$$

When no interference is present ($C_I = 0$), (4.4) reduces to

$$\rho_c = 2T\frac{C}{N_0}\int_{-\infty}^{\infty}\left|H_R(f)\right|^2 G_S(f)\,df \quad (4.5)$$

As stated in [1] and [2], the effective C/N_0 is defined as the carrier–to–noise density ratio (with no interference and only white noise) that would yield the same output SNIR. Thus, according to (4.5), the effective C/N_0 becomes

$$\left(\frac{C}{N_0}\right)_{\text{eff}} = \frac{\rho_c}{2T\int_{-\infty}^{\infty}\left|H_R(f)\right|^2 G_S(f)\,df}$$
$$= \frac{C\int_{-\infty}^{\infty}\left|H_R(f)\right|^2 G_S(f)\,df}{N_0\int_{-\infty}^{\infty}\left|H_R(f)\right|^2 G_S(f)\,df + C_I\int_{-\infty}^{\infty}\left|H_R(f)\right|^2 G_I(f)G_S(f)\,df} \tag{4.6}$$

When the front-end bandwidth β_r is wide enough to receive all of the GNSS signal power, (4.6) can be approximated to

$$\left(\frac{C_s}{N_0}\right)_{\text{eff}} = \frac{C}{N_0 + C_I\int_{-\infty}^{\infty}\left|H_R(f)\right|^2 G_I(f)G_S(f)\,df} \tag{4.7}$$

since $\int_{-\infty}^{+\infty} G_S(f)\,df = 1$. This equation shows that only the interference within the front-end bandwidth causes degradation on C/N_0.

Figure 4.1 shows the effective C/N_0computed for a GPS L1 signal according to (4.7) and the measured C/N_0 by means of a software GNSS receiver in the presence of narrowband (NB) interference (120-kHz width and with a carrier frequency 200 kHz far from the intermediate frequency) and for a different level of interference–to–signal power ratio (C_I/C). The theoretical black line fits the C/N_0 measured by the software receiver thus confirming that (4.7) is a reliable model for estimating the effect of the presence of narrowband interference within the received GNSS frequency band.

The expression in (4.7) is a generalized model for the C/N_0 loss in the presence of NB interference. As we will see in Section 4.3, where the spectral separation coefficient is introduced, such a model cannot be considered valid in the presence of spectrum line interference or pulsed interference.

4.2.1 Theoretical Pulse Blanking Impact on C/N_0 Degradation: Pulsed Interference

In Chapter 2 several pulsed interference sources were introduced. In-band pulsed interference is typical of the aviation scenario due to the presence of other ARNS systems, such as DME or the military TACAN, which uses frequencies located within the L5/E5 bands. In such a scenario, the most common interference mitigation algorithm is the so-called digital pulse blanking. This simple countermeasure is based on the use of digital circuitry that suppresses

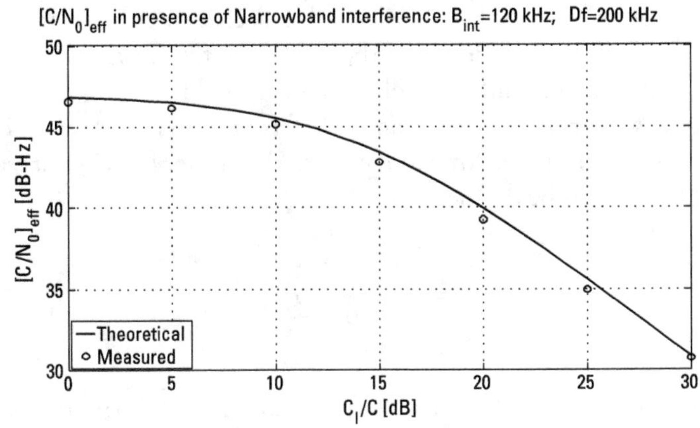

Figure 4.1 Theoretical versus measured effective C/N_0 in the presence of narrowband interference for different C_I/C power ratios.

portions of the incoming signal exceeding a determined blanking threshold. The benefits and drawbacks of this technique will be discussed further in Chapter 6.

Although such a simple strategy is in charge of removing the presence of strong pulsed interference, the blanking operation also leads to a suppression of useful signal, thus impacting the effective C/N_0. The effect of the pulse blanking in the presence of strong pulsed interference has been widely investigated in the literature. A very general expression of the effective C/N_0 in the presence of pulsed interference for a receiver equipped with pulse blanking was first proposed in [3] and then presented in [4] as

$$\left(\frac{C_s}{N_0}\right)_{\text{eff}} = \frac{C}{N_0} \cdot \frac{(1-\beta)}{1+\dfrac{I_0}{N_0}+R_I} \tag{4.8}$$

where β, denoted as the blanker duty cycle, is the total mean activation time of the blanker, which provides the percentage of suppressed signal over the total received signal. The component I_0 takes into account the presence of additional nonpulsed interference, not totally suppressed by the blanker. The term R_I is the aggregate postcorrelation ratio between the residual power of the interfering signal after the blanker and the received thermal noise power. Strong pulses (with peak power over the blanking threshold) as well as weak pulses contribute to the increase in the component R_I. For example, the typical

Gaussian shape of the DME/TACAN pulse, seen earlier in Figure 2.4, and the presence of the modulation over the pulse duration lead to some samples of the pulse itself being under the blanking threshold.

In some scenarios more than one interfering source is present, thus forcing the blanker to be active for a longer period of time. In general, according to [4] and [5], R_I can be defined as

$$R_I = \frac{1}{N_0 \cdot \beta} \cdot \sum_{i=1}^{N} P_i d_i \tag{4.9}$$

where N is the total number of interfering sources, P_i is the received peak power of the ith RFI pulsed signal source, and d_i is the duty cycle of the ith signal source without any pulse collision. More details on the derivation of (4.9) can be found in [5].

Concerning the blanker duty cycle β, a very detailed theoretical derivation in the presence of multiple DME/TACAN sources is presented in [6], where possible pulse superposition is also taken into account.

4.3 Spectral Separation Coefficient

Equation (4.7) provides a theoretical expression for the degradation of C/N_0 in the presence of narrowband interference. In such an expression it can be observed that the impact on C/N_0 also depends on how much the spectra of the received GNSS signal and of the interfering signal overlap. In fact, the component

$$\kappa_I = \int_{-\infty}^{\infty} \left| H_R(f) \right|^2 G_I(f) G_S(f)\, df \tag{4.10}$$

defined as the spectral separation coefficient (SSC), provides a measurement of the overlapping between the received GNSS and the interference power spectral density, which determines the degradation on the C/N_0. In fact, the more the interference components overlap the signal components, the higher the degradation of C/N_0.

The SSC defined in (4.10) takes into account the spectral overlap of the signals, measuring an upper bound to the loss in C/N_0 due to the interference source.

The SSC as defined in [1] and [2] considers the envelope of the power spectral density of the signals. However, due to the almost periodic nature of the GNSS signal, the spectrum has components at frequencies that are a multiple of the inverse of the code period. The impact of the interfering source then also depends on how the signal overlaps with such components, and the SSC value is an upper bound. Furthermore, the estimation provided by the SSC is valid if the signal and interferer can be modeled as statistically stationary and Gaussian. In the case of the GNSS signal, this is generally true if long spreading codes are used. Interference analysis based on this SSC formula does not necessarily hold for signals with short spreading codes for which the spectral "lines" are widely spaced.

The SSC has been used for the assessment of intrasystem and intersystem interference, that is, respectively, interference between signals from different satellites within only one navigation system and interference between signals from one or more different systems [7]. In the latter case, there might also be some degree of code orthogonality between the GNSS signals belonging to different systems, thus further reducing the actual impact.

For the sake of an example, Table 4.1 reports the simulated values of the SSC for the interference of different modulation schemes currently employed in GNSS systems.

Eventually, it has also to be remarked that the definition as such of the SSC does not work in the case of CWI or pulsed interference, due to the line-spectrum nature of their frequency representation.

Figure 4.2 shows the behavior of the SSC for a GPS L1 code signal with respect to the bandwidth of the interfering signal B_i and the interference-to-GNSS frequency offset $\Delta f_i = |f_i - f_{\text{GNSS}}|$, where f_i is the interference carrier and f_{GNSS} is the carrier of the GNSS signal considered.

Table 4.1
Simulated Values of the SSC (in dB) for Different Modulation Schemes

	BPSK(1)	BPSK(10)	BOC(10,5)	BOC(1,1)	MBOC	BOC_{cos}(15,2.5)
BPSK(1)	–61.9	–70.5	–86.5	–67.9	–68.3	–97.4
BPSK(10)		–71.9	–80.9	–70.6	–70.9	–85.0
BOC(10,5)			–73.1	–83.1	–82.8	–88.2
BOC(1,1)				–64.9	–65.3	–92.6
MBOC					–65.0	–92.4
BOC_{cos}(15,2.5)						–70.6

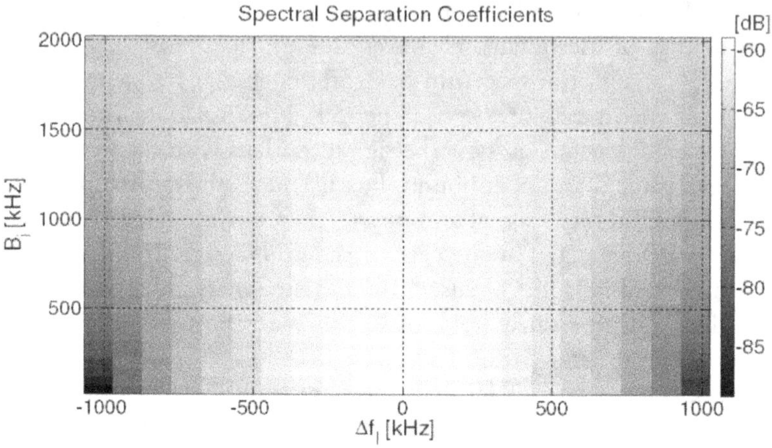

Figure 4.2 SSC values (BPSK (1) signal) versus interference bandwidth B_i and interference-to-GNSS frequency offset Δf_i.

As expected, the SSC is maximum when the two signals are overlapped (small Δf_i), whereas it decreases for larger offsets. Note also that for high values of B_i, the SSC has almost the same value for all of the considered frequency offsets. In these situations, in fact, the interference always overlaps the main lobe of the GNSS spectrum.

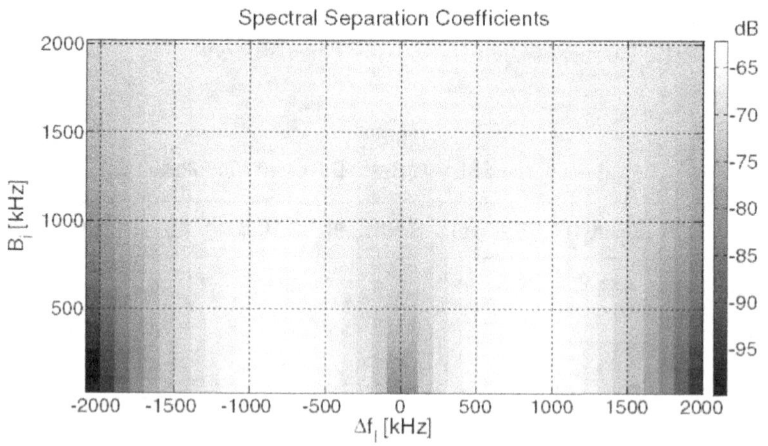

Figure 4.3 SSC values (CBOC (6,1,1/11) signal versus interference bandwidth B_i and interference-to-GNSS frequency offset Δf_i.

For the case of a CBOC(6,1,1/11) employed by the open service of the Galileo system, the SSC resembles the different distribution of the power spectral density, thus reaching maximum values at frequencies shifted by about 1 MHz. The SSC behavior in the case of CBOC(6,1,1/11) is shown in Figure 4.3.

4.4 The Interference Error Envelope

A powerful interference assessment tool able to take into account not only the spectral overlap but the whole effect in the receiver is defined in [8]. As described and shown in Chapter 2, in certain cases the power level of an interfering signal does not totally block the receiver; it does, however, degrade receiver performance in terms of position accuracy.

The interference error envelope (IEE) was proposed in [8] as an evolution of the SSC, able to take into account also the effect of the receiver architecture in the evaluation of the interference impact.

The IEE is defined by a procedure that is similar to the one proposed to assess the effect of the multipath, and it aims to assess the bias induced on the discrimination function of the receiver by the presence of interference. It is defined as the measure of the maximum discriminator function distortion (i.e., bias of the stable tracking point) with respect to one (or more) parameters of the interfering signal: the ranging error (expressed in meters) is plotted versus one or more variable interference characteristics (e.g., the carrier frequency for CWI).

The bias can be theoretically defined as

$$b_{\max}(f_i) = \alpha \frac{2}{NL} \int_{-\infty}^{\infty} |I(f)||W(f)||C(f)| \sin(\pi f \Delta)\, df \tag{4.11}$$

where:

$|I(f)|$ = the discrete time Fourier Transform (DTFT) of the sampled interfering signal;

$|W(f)|$ = the DTFT of the digital impulse response modeling the front-end;

$|C(f)|$ = the DTFT of the sampled ranging code signal;

NL = the total number of samples in the integration time, with L being the number of samples in one code period and N the number of integrated code periods.

The parameter α is related to the slope of the interference-free discrimination function

$$\alpha = -\frac{cT_c\Delta}{2} \tag{4.12}$$

where c is the speed of light, T_c is the chip period, and Δ is the early-late correlator spacing.

For a CWI with amplitude A and carrier f_i, (4.11) is reduced to

$$b_{\max}^{\mathrm{CW}}\left(f_i\right) = \alpha\frac{2A}{NL}\left|W\left(f_i\right)\right|\left|C\left(f_i\right)\right|\sin\left(\pi f_i\Delta\right) \tag{4.13}$$

where $b_{\max}(f_i)$ is composed of three main components that depend on the (1) code elementary function, (2) code line spectrum, and (3) discriminator spacing Δ. In fact, the contribution $C(f_i)$ depends on the modulation (the elementary pulse shaping the chip) and on the specific code sequence.

Figure 4.4 plots the different partial contributions to $b_{\max}(f_i)$ versus the frequency shift $\Delta f_i = \left|f_i - f_{\mathrm{GNSS}}\right|$, where f_{GNSS} is the carrier of the GNSS signal considered. The IEE then allows us to take into account the model of the receiver as well as the effect of the GNSS signal shape and it allows prediction of the bias induced by the CWI.

Furthermore, Motella et al. [8] also demonstrated how the behavior in the case of narrowband interference (NBI) or wideband interference (WBI) can be inferred starting from the IEE obtained for a CWI. The bias induced by WBI (or NBI) can be obtained by applying a moving average filtering of the CW IEE, such as

$$b_{\max}^{\mathrm{WB}}(f_i) \approx \frac{1}{M}\sum_{k=i-M/2}^{i+M/2} b_{\max}^{\mathrm{CW}}\left(f_k\right) \tag{4.14}$$

where M is the number of frequencies of the CWI IEE corresponding to the bandwidth of the NBI or WBI.

The IEE is then also a useful tool for the design of the receivers, since it allows designers to choose a receiver setup that is robust to interfering signals that are likely to appear in the operating environment at specific frequency shifts with respect to the carrier frequency.

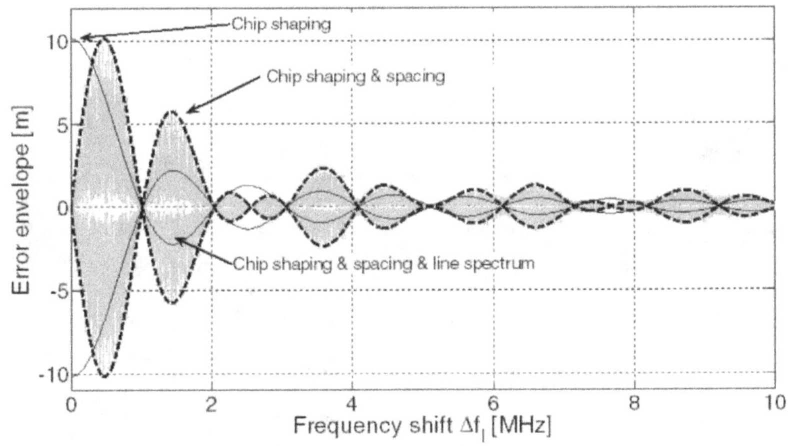

Figure 4.4 Theoretical contribution effects on the IEE; Δ = 0.2 chip, BPSK (GPS L1, PRN 1) signal, CW interference, CW frequency step = 1 kHz, BW = 30 MHz.

In [8] several cases of real and simulated signals are presented, validating the theoretical derivation of the IEE. The $b_{max}(f_i)$ is the envelope of the values that can be obtained by real measurement or by simulation. Examples of the CWI IEE for different modulation schemes employed in GNSS systems are depicted in Figure 4.5.

When the IEE is used for comparison among different case studies, it may not be practical for the identification of the optimal case, due to the trend of the plots to not be smooth. In [8] the authors propose to use the interference running average (IRA) defined as

$$\mathrm{IRA(B)} = \frac{1}{B}\sum_{0}^{B} \frac{\mathrm{IEE}_{\max}(f) + \left|\mathrm{IEE}_{\min}(f)\right|}{2} df \tag{4.15}$$

This curve represents the potential impact of interference, depicting for each frequency value B the cumulative worst case error averaged over the bandwidth between 0 and B.

The interference running average curves are depicted in Figure 4.6 for different modulation schemes, in the case of $\Delta = 0.2$ chip. Using the IRA representation it can be easily noted at which frequency shift f_i the interfering signal has the largest impact.

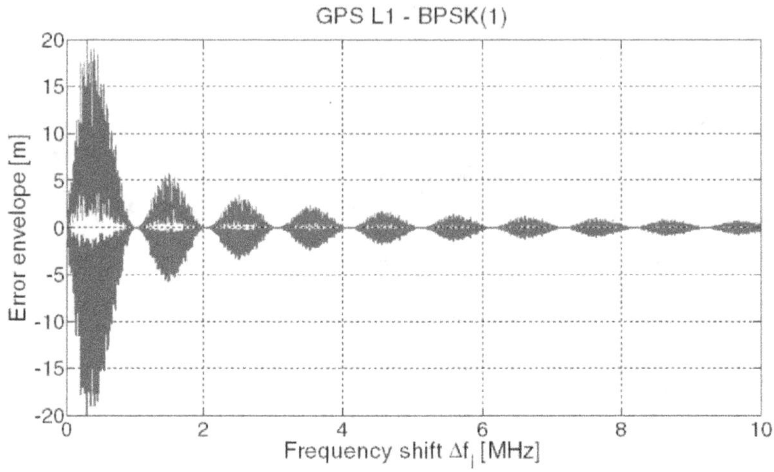

(a)

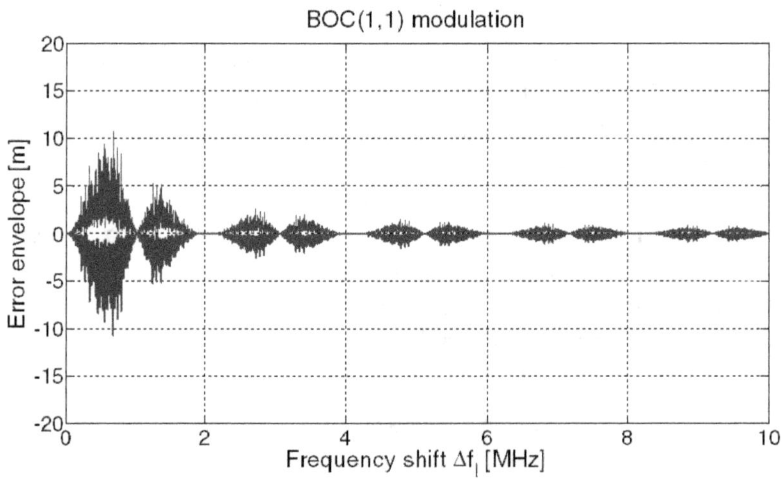

(d)

Figure 4.5 CWI IEE for different modulation schemes: (a) BPSK(1), (b) BOC(1,1).

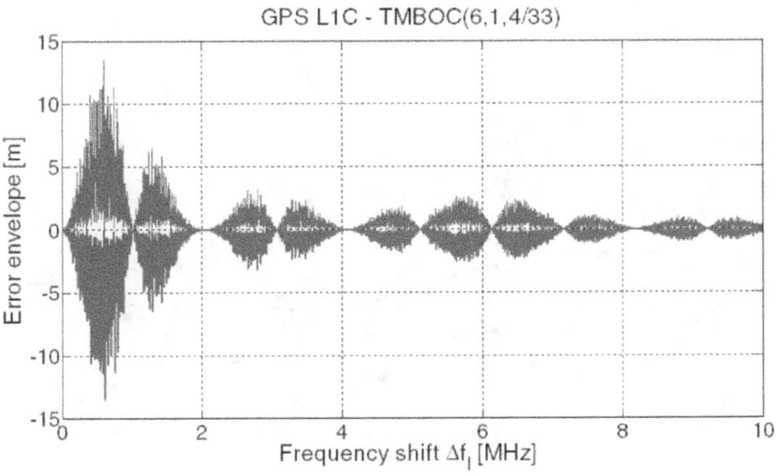

(c)

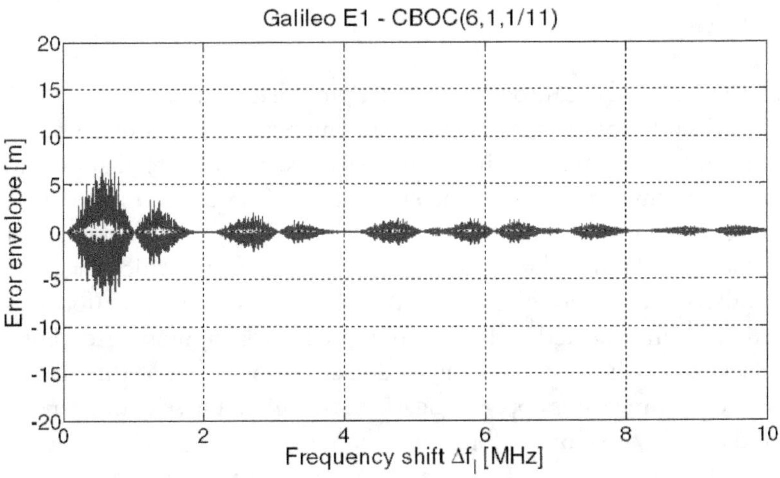

(d)

Figure 4.5 *(Continued)* (c) TMBOC(6,1,4/33), and (d) CBOC(6,1,1/11) for $\Delta = 0.2$.

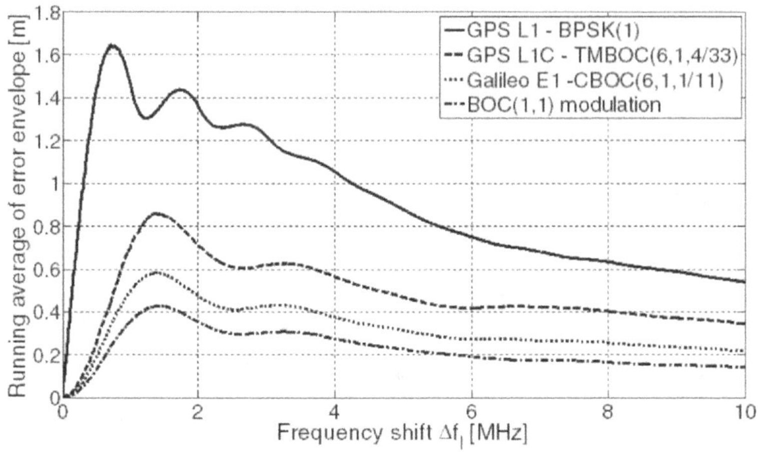

Figure 4.6 Interference running average comparison of different modulated signals [BPSK(1), BOC(1,1), CBOC(6,1,1/11), and TMBOC(6,1,4/33)] in the presence of CWI, Δ = 0.2 chip.

4.5 Conclusions

In this chapter analytical tools have been provided for the assessment of the impact of interfering sources on the performance of GNSS receivers. They might be useful during the design phase of the receivers if information is available about the environment in which the receiver is expected to operate. This is the case, for example, for GNSS receivers that are expected to operate in the ARNS bands and tailored to aeronautical applications, for which the presence of the pulsed interference of DME/TACAN stations is a fact. Unfortunately, interference is unexpected and it is also unpredictable in most cases. For such a reason receivers have to be ready to detect the presence of the interfering source and in some cases apply mitigation strategies. These countermeasures will be the main topic of the following chapters.

References

[1] Betz, J. W., "Effect of Narrowband Interference on GPS Code Tracking Accuracy," *Proc. 2000 National Technical Meeting of the Institute of Navigation*, Anaheim, CA, January 2000, pp. 16–27.

[2] Betz, J. W., "Effect of Partial-Band Interference on Receiver Estimation of C/N_0: Theory," *Proc. 2001 National Technical Meeting of the Institute of Navigation*, Long Beach, CA, January 2001, pp. 817–828.

[3] Radio Technical Commission for Aeronautics, "Assessment of Radio Frequency Interference Relevant to the GNSS L5/E5a Frequency Band," Technical Report RTCA DO-292, 2004.

[4] Erlandson, R. J., et al., "Pulsed RFI Effects on Aviation Operation Using GPS L5," *Proc. 2004 National Technical Meeting of the Institute of Navigation*, San Diego, CA, January 26–28, 2004, pp. 1063–1076.

[5] Musumeci, L., J. Samson, and F. Dovis, "Performance Assessment of Pulse Blanking Mitigation in Presence of Multiple Distance Measuring Equipment/Tactical Air Navigation Interference on Global Navigation Satellite Systems Signals," *IET Radar, Sonar & Navigation*, IET, Vol. 8, No. 6, pp. 647–657, July, 2014. doi:10.1049/iet-rsn.2013.0198

[6] Bastide, F., et al., "GPS L5 and GALILEO E5a/E5b Signal-to-Noise Density Ratio Degradation due to DME/TACAN Signals: Simulations and Theoretical Derivations," *Proc. 2004 National Technical Meeting of the Institute of Navigation*, San Diego, CA, January 26–28, 2004, pp. 1049–1062.

[7] Titus, L. B. M., et al., "Intersystem and Intrasystem Interference Analysis Methodology," *Proc. ION GPS/GNSS 2003*, Portland, OR, September 2003, pp. 2061–2069.

[8] Motella, B., et al., "Method for Assessing the Interference Impact on GNSS Receivers," *IEEE Trans. on Aerospace and Electronic Systems*, Vol. 47, No. 2, 2011, pp. 1416–1432.

5

Interference Detection Strategies

Emanuela Falletti and Beatrice Motella

5.1 Introduction

The extremely low power at which a GNSS signal-in-space (SIS) is received with conventional hardware makes GNSS vulnerable to interference from other systems, whose sources are ground based and therefore far closer to the receiver than the GNSS satellites (i.e., the *near-far effect*).

The various mechanisms through which interference attacks the normal functioning of a GNSS receiver can be recognized among the following [1]:

1. Saturation and desensitization of front-end low-noise amplifiers, mixers, and other circuitry;
2. Reciprocal mixing effects that arise from the fact that receivers cannot generate a perfect tone to downconvert the desired signals;
3. Intermodulation products;
4. Aliasing of out-of-band emissions that remain after filtering into the receiver's passband;
5. Reception of in-band (to GNSS) emissions that are always present due to imperfections in the signal generation and filtering of the interfering system.

In this classification as well as in the rest of this chapter, intentional attacks to the SIS structure, such as spoofing and meaconing mechanisms, are not considered, because they are intrinsically undetectable by generic anti-interference techniques, which assume to cope with "unstructured" interference signals.

In all the other cases, radio-frequency interference (RFI) detection techniques aim at recognizing the "deviations from the normal conditions" (distortions) caused by the presence of interference impinging on the GNSS antenna, in order to raise timely alerts or activate the proper mitigation technique.

Because of the spread-spectrum nature of GNSS signals, there are situations in which narrowband interference, although clearly present in the received signal spectrum, cannot impact the clean processing of the SIS [2]; for example, this is the case for continuous-wave (CW) interference whose carrier frequency falls between two GPS C/A code lines, so that the undesired signal is completely filtered out by the correlation process. In contrast, an intentional jamming attack or a very strong unintentional source might be such as to completely saturate the initial stages of the RF front-end and consequently to obscure any SIS signal reception [3].

Between these two borderline cases, the impact of interference is extremely various and might include situations in which the interferer partially degrades the regular signal processing at different stages of a conventional GNSS receiver's chain, as presented, for example, in [4] and discussed in Chapter 2 of this book. Nonetheless, the impact of interference could be attenuated by incorporating digital signal processing techniques to the GNSS receiver and by adequately designing the receiver analog front-end (FE).

A huge variety of techniques for interference detection have been proposed in the literature. These techniques exploit different receiver observables taken at different stages of the receiver signal processing chain and leverage specific properties of such observables. A special family of techniques is based on *spatial diversity*, which effectively copes with the interference by means of opportune steering or shaping of the antenna beam pattern. This approach requires the presence of an antenna array, with digital or analog steering capabilities (controlled radiation pattern antenna (CRPA)), and is beyond the scope of this book. The interested reader may refer to [5, 6] and references therein.

In light of this, interference detection techniques at the receiver stage can be broadly classified as follows:

1. Interference detection via AGC monitoring;
2. Interference detection via time-domain statistical analysis;
3. Interference detection via spectral monitoring;
4. Interference detection via postcorrelation statistical analysis;

5. Interference detection via carrier–to–noise power ratio monitoring;
6. Interference detection via pseudorange monitoring;
7. Interference detection via PVT solution observation.

Notice that techniques 1, 2, and 3 are *precorrelation* techniques, meaning that they apply to the spread-spectrum received signal before any despreading operation (correlation with a local code). This fact implies that precorrelation techniques are potentially able to warn the receiver managing procedures about nonharmful interference. On the other hand, the *postcorrelation techniques*, items 4 to 7, require the preexisting condition of normal signal acquisition and tracking to be met and could, therefore, be less effective in cases of strong interfering power.

Figure 5.1 shows a conceptual block diagram of the signal processing stages of a GNSS signal receiving chain, where the observables necessary to implement the various interference detection techniques are highlighted. The following sections offer an overview of each of the families of detection techniques listed above.

A word of caution is worthwhile before discussing the details of the various techniques: Unwanted interference is by nature an unpredictable event of unknown onset, duration, and temporal and spectral content. However, in general, detection techniques can be tailored to specific interference

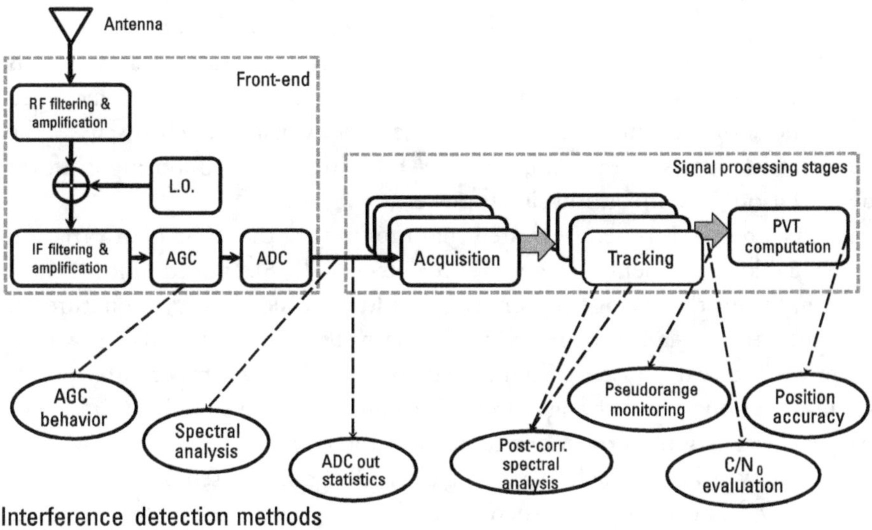

Figure 5.1 Conceptual diagram of the GNSS signal processing blocks. The observables relevant for different interference detection methods are highlighted.

characteristics only, typically by trading off between sensitivity with respect to a specific interference type and generality over all the possible types. The implementation complexity of each technique is another issue that limits the applicability to certain classes of receivers only. For these reasons, interference monitoring must be a mixture of techniques, whose overall complexity and performance is limited by the processing capability allocated to these functionalities by the receiver design.

5.2 Interference Detection via AGC Monitoring

Modern GNSS receivers are composed of an analog front-end and a digital part for SIS processing (signal acquisition and tracking, data demodulation and pseudorange computation, position–velocity–time solution computation). Before the analog-to-digital converter (ADC), the automatic gain control (AGC) mechanism acts as an adaptive variable gain amplifier, whose main role is to minimize the quantization losses by adjusting the received signal level to the ADC input range, under the assumption of certain amplitude probability density distributions (see Figure 5.2). For GNSS receivers, in which the useful signal power is below that of the thermal noise floor, the AGC is driven by the noise environment rather than the signal power. In the presence of strong interfering signals (that might saturate the quantization process), the AGC may increase the dynamic range, decreasing its gain and limiting signal saturation. The AGC response to the interference is not linear and strongly depends on its design in terms of sensitivity and reactivity; it is typically a piecewise constant function of the analog input power, averaged along the designed time interval. Reference [7] very instructively discusses the quantization losses as a function of the ADC thresholds (supposing uniform quantization) and incoming signal power (or variance).

As a result, it is demonstrated that monitoring the AGC gain variation is a valuable instrument for interference detection [7, 8]. Indeed, under nominal conditions the average power level should be of slowly varying nature, so that the variable gain is expected to be relatively stable and confined within a known interval (noise and gain variations caused by temperature, power supply, environmental changes around the antenna, and so on), so sudden or strong variations are indications of possible interference.

The typical approach to steering the AGC gain is based on the measure of the ADC output. In [9] a theoretical discussion on the estimation of the ADC input power on the basis of the ADC output power is given. So AGC steering can be performed with the estimated ADC output power/variance put

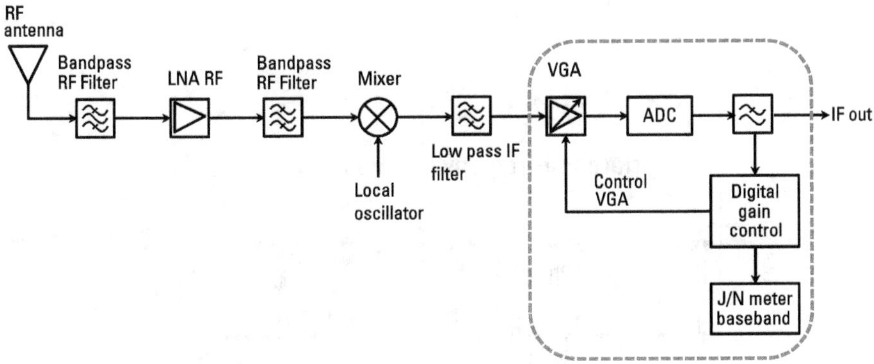

Figure 5.2 RF front-end block diagram. The dashed box emphasizes the AGC/ADC functional block with digital feedback gain control. Although other implementations are possible, this scheme has a digital implementation that uses ADC output samples to form the metric used to steer the AGC gain.

at the input of a proportional and integral (PI) controller or a look-up table, which is also the most common AGC implementation in a GNSS receiver [10].

The AGC monitoring technique to detect interfering signals has been recently adopted to survey the GPS L1/Galileo E1 bands in proximity to airports [11]. It was also used in [4, 12] for the detection of interference from a TV transmitter station. The authors of [4] were able to detect the effect of the third harmonic emission of a DVB-T station within the L1 GPS bandwidth, using a low-cost GNSS front-end and a software receiver. The signal spectrum at the front-end output was distorted by the presence of several narrowband and wideband interfering signals. Comparing the trend of the interfering signal power with the AGC level, it was easy to observe how the adaptive gain was driven by the presence of the interference: when the interference was present, the AGC avoided the ADC saturation, decreasing the applied signal gain. Figure 5.3 shows 1 minute of the variable gain measurements taken in the mentioned experiment, compared with 1 minute of measurements taken in a controlled interference-free environment; clearly, the power fluctuations due to the interference force the AGC to continuously readapt the gain at the ADC input, while in the absence of interference it remains constant.

5.2.1 The Role of the ADC

Note that the ADC is a key element of an interference-resistant FE. Typical commercial GNSS receivers for consumer applications have low-bit-count ADCs (typically, 1 to 3 quantization bits), but such a configuration is prone

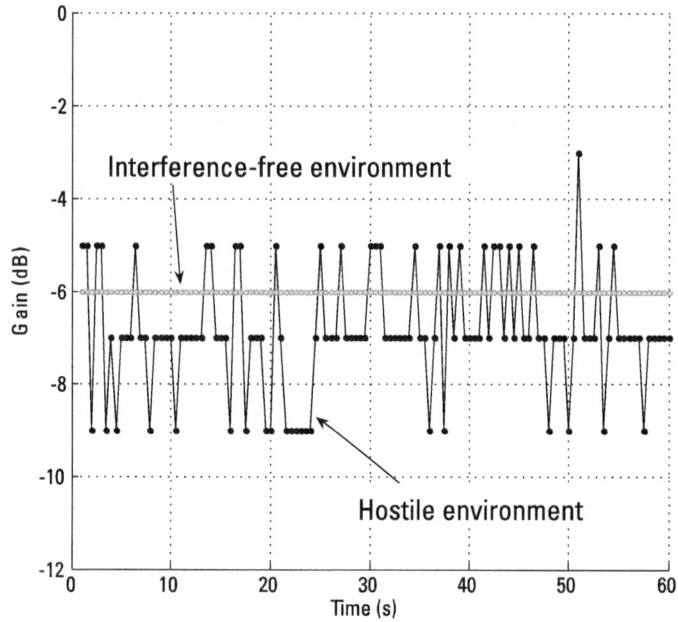

Figure 5.3 Experiment of adaptive gain behavior at the ADC input. In an interference-free environment, the gain introduced by the AGC is kept constant over time at –6 dB. The adaptive gain behavior changes in an interfered environment; it is not constant and varies within a range of 6 dB, with an average level lower than in the previous case.

to frequent SIS signal quantization losses when interference passes through the analog FE stages to the AGC input [7]. Indeed, in the presence of powerful interference, the variable gain applied to the signal at the ADC input would reduce the total input power so as to optimize the use of the input dynamic range; however, the strong attenuation applied to the received signal would also strongly reduce the amplitude of the SIS component below the resolution of the least significant bit at the ADC output, thus preventing any possibility of recovering the digitized SIS signal content after despreading.

For this reason, whenever a receiver is expected to implement advanced techniques for interference detection and mitigation, a good practice could be to allow at least 2 to 3 quantization bits more than the number required in the signal acquisition and tracking stages [7, 13]. In normal (i.e., interference-free) conditions, the variable signal gain should be driven so as to avoid the use of these "spare" bits. They will be used only when a signal is received with interference in order to minimize quantization losses on the SIS. This approach is particularly suitable when associated with digital signal blanking to mitigate pulsed interference. A more detailed discussion of this topic is presented in Chapter 6.

5.3 Interference Detection via Time-Domain Statistical Analysis

Time-varying interference can be identified by applying statistical analysis techniques based on the observation of the time fluctuations of the signal sample distribution at the ADC output [7, 14–16]. The techniques are based on the fact that the data to be processed (i.e., the signal samples at the output of the front-end ADC) can be modeled as random processes, whose "statistical characteristics" are sensitive to the presence of interference, and then might be monitored for detection purposes. These types of methods are widely used in disciplines such as economics and biology, but examples in the GNSS field, while promising, are still rare.

Since the spectral and statistical characteristics of the interference are unknown, a nonparametric goodness-of-fit (GoF) detection method is proposed and analyzed in detail in [15] and [16], where it is associated with a chi-square test to reduce the dimensionality of the problem [17].

For interference detection purposes, the IF samples at the output of the ADC can be used to build the test statistic and make the decision. In the presence of an interference source, the regular shape of the histogram will be modified, enabling the test to detect such a distortion. The objective is to decide on the presence of a certain signal component (the interference), based on the measurement of a finite sequence of samples (the signal samples at the ADC output). This is a typical problem of detection that can be formulated in terms of hypothesis testing, where a binary hypothesis can be stated as follows

$$H_0 \text{ (RFI absent)} \quad : \quad p_X(x) = p_Y(x)$$

$$H_1 \text{ (RFI present)} \quad : \quad p_X(x) \neq p_Y(x)$$

where $p_Y(x)$ and $p_X(x)$ are first-order probability density functions (PDFs) of a stationary random process (at least stationary in the observed time interval, i.e., $0 \leq n < N$). In our case the processes $X[n]$ and $Y[n]$ represent the received signal (downconverted and quantized) in the presence of interference and in its absence, respectively. Moreover, $Y[n]$, since it is dominated by the noise, can be modeled as a zero-mean white Gaussian process. The knowledge of the process distribution where there are no interfering signals (when H_0 is verified) is the only requirement posed by the chi-square GoF, which does not need any other information on the interference characteristics (when H_1 in verified). The method works as indicated by the following points:

1. First, the discrete version of the PDF of $X[n]$ must be evaluated when the H_0 hypothesis is verified. The method takes a set of measurements $\mathbf{x}_m = \{x_m[0], x_m[1], \ldots, x_m[N-1]\}$ to build the reference PDF in the form of a reference histogram $\mathbf{E} = \{E_1, E_2, \ldots, E_k\}$, where k is the number of bins and $\mathbf{x}_m$ can be seen as an instance of a random vector $\mathbf{X} = \{X[0], X[1], \ldots, X[N-1]\}$, where N is the number of observed data.
2. The method takes a set of measurements $\mathbf{x}_m = \{x_m[0], x_m[1], \ldots, x_m[N-1]\}$ (this time the interference signals might be present) and the values are grouped and counted for each bin to form the vector $\mathbf{O} = \{O_1, O_2, \ldots, O_k\}$. At this point, two histograms are available: the reference histogram, $\mathbf{E}$, and the observed histogram, $\mathbf{O}$, representing $p_X(x)$ and $p_Y(x)$, respectively.
3. The method evaluates the test statistic

$$T_\chi(\mathbf{x}_m) = \sum_{i=1}^{k} \frac{(\mathbf{O}_i - \mathbf{E}_i)^2}{\mathbf{E}_i}$$

The value of $T_\chi(\mathbf{x}_m)$ is used to discriminate between the two hypotheses H_0 and H_1. When the two histograms (reference and observed) perfectly coincide, $T_\chi(\mathbf{x}_m) = 0$. Intuitively, we can say that the higher the value of the test statistic is, the less similar the two histograms will be. A proper threshold has to be set to result in a decision between H_0 and H_1. The test statistic $T_\chi(\mathbf{x}_m)$ can be seen as an instance of a random variable $T_\chi(\mathbf{x})$, and it is possible to demonstrate that, for large N, the variable $T_\chi(\mathbf{x})$ is approximately χ^2 distributed with $k-1$ degrees of freedom.
4. The probability

$$p_m = \Pr\{T_\chi(\mathbf{x}) > T_\chi(\mathbf{x}_m)\}$$

referred to as the *p-value*, is evaluated. We observe that $p_m \simeq 1$ means $T_\chi(\mathbf{x}_m) \simeq 0$, therefore, the two histograms are almost identical. In contrast, when $p_m \simeq 0$, the two distributions are different. The decision is therefore made by fixing a threshold p_α, known as the *level of significance*, as follows:

$$p_m > p_\alpha \quad : \quad H_0 \text{ is accepted}$$

$$p_m < p_\alpha \quad : \quad H_0 \text{ is rejected}$$

For further details on the procedure, refer to [16].

Figure 5.4, from [15], reports an example of detection of an interfering sinusoidal signal with variable frequency that appeared with an intermitting pattern along 2.5s of data. It is easy to observe how the identification of the presence of the interference would not be straightforward in either the time or frequency domains [Figures 5.4(a) and (b)]. In contrast, Figure 5.4(c) shows that the chi-square GoF test profiles clearly discriminate the cases of the presence and absence of interference, enabling a threshold-based detection mechanism.

The major advantage of such a method is the fact that it is applicable to all types of interference and the required computational effort is affordable. Furthermore, the method is able to detect the presence of interference signals also at very low power levels.

5.4 Interference Detection via Spectral Monitoring

Because in normal conditions the received SIS power level is below the thermal noise power level at the antenna (assumed white over the whole digitization bandwidth), a spectral estimate of the received signal is expected to redraw the equivalent transfer function of the front-end, multiplied by the noise variance that passed through the analog front-end. The interference-free expected spectral estimate can be assumed known after proper calibration of the antenna and RF front-end setup. Indeed, relatively small variations of the average received power can be expected as a function of temperature and environmental changes around the antenna, and they should be further limited in the presence of AGC.

In these conditions, an interfering signal impinging on the antenna with the power level exceeding the noise floor is expected to be detectable via spectral analysis by comparing the estimated power spectral density of the received signal with a spectral mask that appropriately represents the nominal interference-free conditions.

Spectral monitoring is naturally associated with frequency-domain interference suppression techniques, such as frequency excision [18] and

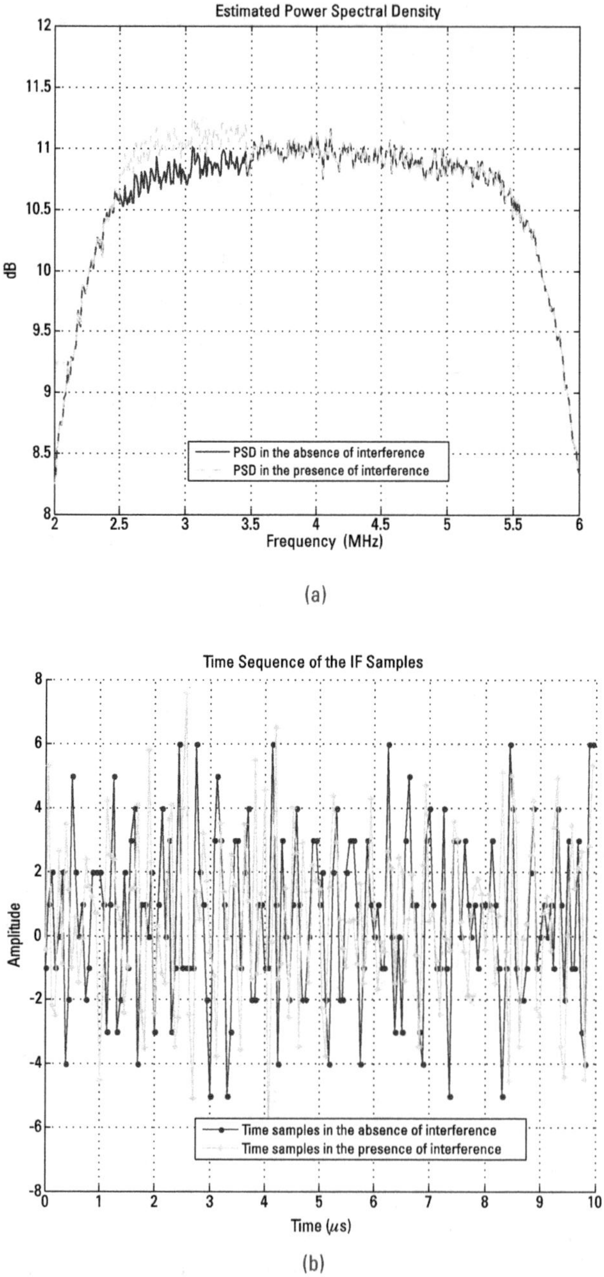

Figure 5.4 Results of the chi-square goodness-of-fit test: (a) spectrum of the IF signal in the presence and absence of interference; (b) time samples of the IF signal in the presence and absence of interference.

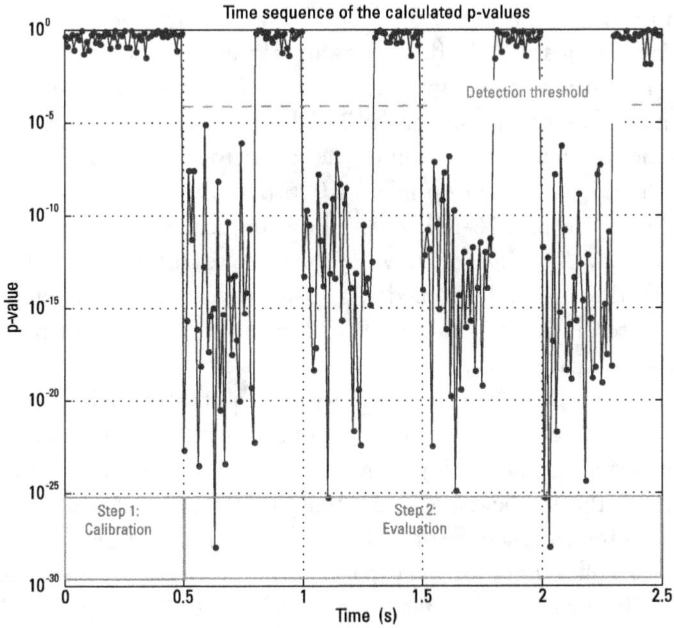

(c)

Figure 5.4 *(Continued)* (c) profile of the detection metric of the chi-square GoF test.

time-frequency excision [19–21]. It can also be used to drive the activation of notch filters to suppress narrowband interference. Notch filters, which are currently implemented in many commercial receivers, will be discussed in Chapter 6.

Basic spectral estimation can be implemented via simple normalized fast Fourier transform (FFT) or periodogram methods (which are based anyway on the use of sequences of shorter and windowed FFTs) [22, 23]. Such nonparametric spectral monitoring techniques are conceptually simple to implement, but their performance is inherently limited by a set of factors [18, 22–25]: First, they need relatively long observation windows (on the order of several hundreds of milliseconds) to produce spectral estimates with reduced estimation variance; second, periodograms (whichever type is used: sample, Bartlett's, Welch's) are biased estimators, which introduces spectral leakages in correspondence with sharp spectral peaks and nulls; third, they are heavily based on the use of the FFT, which is a demanding resource whose complexity is superlinear with respect to the number of input samples.

As a result, the parameters of the FFT algorithm used in each specific implementation must be carefully chosen, taking into account the necessary frequency resolution, the digitization bandwidth, and the computational resources available to compute each FFT. Indeed, the FFT length is directly related to the frequency resolution of the spectrum, normalized to the whole digitization bandwidth. For example, a 4,096-point FFT ($N_{FFT} = 4096$) applied to a signal sampled at a 16-MHz sampling frequency (f_s) offers a frequency resolution of $f_s/N_{FFT} \approx 3.91$ kHz, while the same resource (in number of FFT points) applied to a signal sampled at a 40-MHz sampling frequency reduces its frequency resolution to $f_s/N_{FFT} \approx 9.77$ kHz and, at the same time, requires 2.5 times faster operation.

Furthermore, the length of the signal observation window (which is defined on the order of one to several times the FFT length, depending on the type of periodogram employed) is a limiting factor of the temporal resolution of such methods, that is, of their ability to detect the starting and ending instants of an interference event. A large observation window leads to better results in terms of event detection capability; nevertheless, it causes a reduction in the accuracy with which a relatively short interference event can be localized in time. A critical trade-off between detection capability and detection reliability has to be accounted for in these cases [23].

Because the FFT-based periodograms do not properly work for irregular short events, in these cases time-frequency analysis techniques can be applied [26]. Several alternative distributions exist (short-time Fourier, Wigner-Ville, Choi-Williams, and so on), but their performance generally depends on the type of interference to be detected [19, 27]. A critical issue with such a family of techniques is the bidimensional search domain (namely, time and frequency), which requires a significant computational burden to be handled.

Another bidimensional search domain is defined by the time-scale analysis techniques, based on the use of the two-dimensional wavelet transform. These techniques are gaining interest in terms of GNSS interference monitoring [28, 29]. Chapter 7 of this book is devoted to these techniques and to the mitigation strategies they enable.

5.5 Interference Detection via Postcorrelation Statistical Analysis

Hazardous types of interference are those that are able to leak through the correlators to degrade the postcorrelation measurements, ultimately causing a degradation of the PVT estimation performance. Thus, unwanted spectral components due to interference could be recognized by applying appropriate detection metrics to the signal at the correlator output. For example, researchers [30, 31] have exploited a parametric spectral analysis based on the principles of the harmonic analysis of random processes to detect the presence of a narrowband signal mixed to the SIS, monitored at the output of a multicorrelator engine. The principle is that the vector of samples (punctual correlations) at the output of a bank of correlators spanning one or more chip intervals, say, $\mathbf{v}[k]$ where k is the time index, can be subjected to harmonic analysis—employing parametric spectral estimators such as the MUSIC method or one of its variants [32]. The largest eigenvalues of the sample covariance matrix of $\mathbf{v}[k]$ are associated with the presence of strong harmonic components in the observed $\mathbf{v}[k]$, and can therefore be used to detect a narrowband interference that affects the correlators output [31], and finally to drive the activation of a proper notch filter.

Figure 5.5 illustrates an example of multicorrelator output, obtained with a software receiver [33]; correlators are spaced $\delta_{MC} = 4$ samples apart, that is, $4/f_s$ s, being $f_s = 13$ MHz the sampling frequency of the front-end.

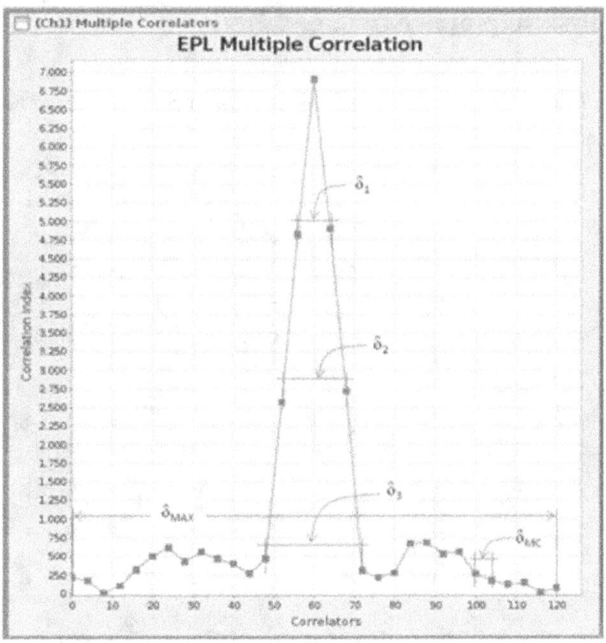

Figure 5.5 Output of a multicorrelator, consisting of a bank of 15 pairs of early–late correlators plus a prompt one.

The multicorrelator spacing δ_{MC} determines the frequency range of detectable interferers in the signal spectrum $(-F_{MAX}, F_{MAX})$ around the IF frequency:

$$F_{MAX} = \frac{1}{2}\frac{1}{\delta_{MC}} = \frac{f_s}{8} \approx 1.6 \text{ MHz}$$

which is enough to detect interfering lines falling within the main spectral lobe of an E1/L1 signal.

The eigenvalues extracted by the eigen decomposition of the sample covariance matrix of $\mathbf{v}[k]$ are shown in Figure 5.6 for both an interference-free and a CW-interfered SIS, with a carrier-to-interference power ratio of $C/I = -20$ dB. When interference is present, the amplitude of the eigenvalues is about 2 orders of magnitude larger than in the interference-free case.

The computational complexity of a multicorrelator bank followed by a harmonic analysis is compensated for by the accuracy in estimating the harmonic components in $\mathbf{v}[k]$, which is known to overcome the accuracy of a non parametric spectral analysis [32].

It is interesting to note that the degradation of correlator output due to an interference component could also be recognized using time-domain statistical analysis, as proved in [15].

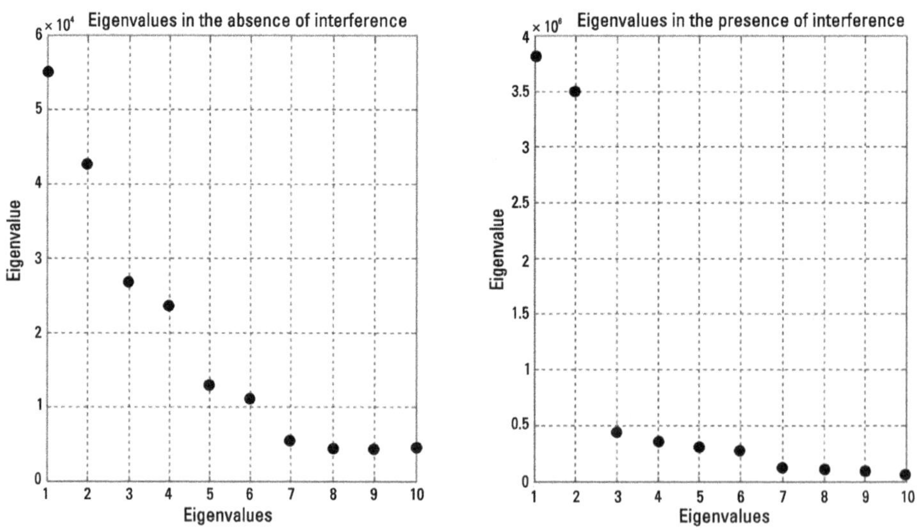

Figure 5.6 Eigenvalues of the covariance matrix of $\mathbf{v}[k]$ in the absence of interference (left) and in the presence of CW interference (right). Notice the different scale of the two plots.

5.6 Interference Detection via Carrier-to-Noise Power Ratio Monitoring

A parameter able to sense a degradation in the quality of the signal tracking is the estimated C/N_0 ratio for each satellite in tracking. It is the estimated ratio between the power associated with the SIS and the power spectral density of the noise at the antenna. A compromised code/carrier tracking determines a reduction in the C/N_0 level computed by the receiver, as it is typically based on postcorrelation observations [34, 35]. Indeed, it is common practice in the receivers to exclude from the set of satellites used for the position–velocity–time (PVT) estimation (satellites *in navigation*) those experiencing a C/N_0 ratio below a certain threshold (e.g., 30 dB-Hz), based on the fact that a low estimated C/N_0 ratio indicates a low-quality tracking condition.

The C/N_0 ratio reduction may be caused by several factors, such as the presence of a non-line-of-sight (NLOS) propagation condition of the SIS, a temporary SIS outage, a significant multipath fading effect, a strong Doppler rate nonperfectly tracked by the carrier tracking loop, the gain of a directional antenna at low elevations, and the presence of interference. In the simplest case, a progressive reduction of the C/N_0 ratio indicates a satellite that is going to disappear behind the horizon, for which the low elevation increases the probability of blocked, NLOS, and faded propagation.

Although the source of the impairment is impossible to discriminate from the sole observation of the C/N_0 ratio, nevertheless it is a powerful indicator of a critical condition occurring to a specific satellite signal.

Therefore, although not a stand-alone detection method, the C/N_0 level observation can be used to *assess the impact* of an interference on the tracking quality of the satellite signals.

The following example attempts to demonstrate the effect of a CW interference on two GPS C/A signals, characterized by two different pseudorandom noise (PRN) codes.

As it is well known, the GPS C/A code set is based on the Gold code characteristics, repeated with a 1-ms period. Such a periodic code has a line spectrum with lines spaced 1 kHz apart. Because of the presence of data modulation, each spectral line is convolved with the narrowband spectrum of the data at a 50-Hz data rate. Depending on the specific code word, some lines are stronger than others. This means that CW interference might mix with a strong C/A code spectral component and leak through the correlator, degrading the correlation quality. Such degradation is witnessed by the C/N_0 measurement, which shows a drop. For a given interfering power, the level of degradation essentially depends on the power associated with the code component subject to the interference. Hereafter, a simple experiment is presented, based on the C/N_0 estimation produced by a software receiver [33]. It compares the C/N_0 degradation caused by two CW interfering signals, which match the strongest and the second strongest component of the GPS C/A PRN 5 code. Two data collections have been considered. In both of them the interference sweeps ±500 Hz around a code line: In the first case around the strongest one (located at L1+23 kHz), and in the second case around the second strongest one (located at L1+199 kHz). The power of the interference was constant at −85 dBm, superimposed to a SIS received at −92 dBm. Figure 5.7 depicts the spectral representation of the test setup.

The result is shown in Figure 5.8, where the C/N_0 estimates are reported for the two test cases, compared with a condition of interference-free signal processing (continuous line). In the first test case (dashed line), the CW interfering line sweeps around the strongest code line from −500 Hz to +500 Hz around L1+23 kHz; the degradation with respect to the interference-free C/N_0 estimate is on the order of 11 dB when the interfering line is "far" from the code component (i.e., at the beginning and the end of the data collection), and drops to more than 26 dB in the middle of the data collection, when the interfering tone matches the code component. This drop corresponds to a frequency interval of ±100 Hz around the main code line. On the other hand,

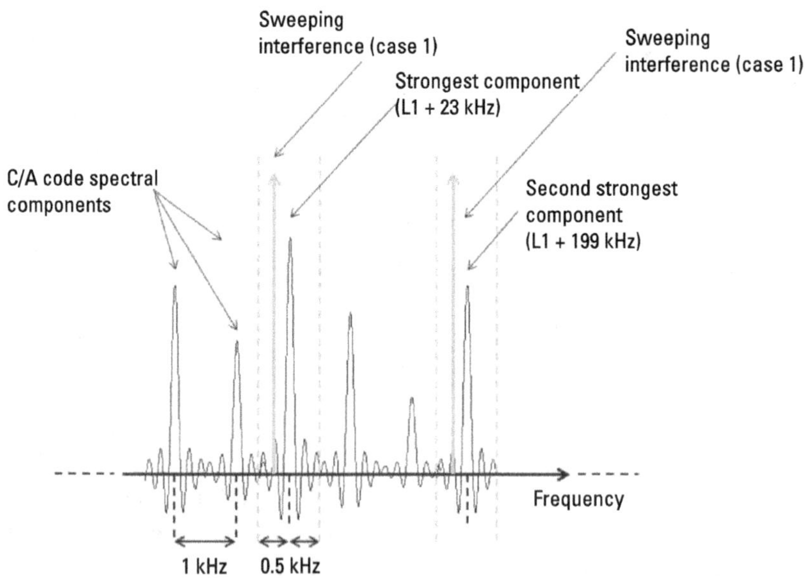

Figure 5.7 Spectral representation of the C/A code components and of the swept CW interferences.

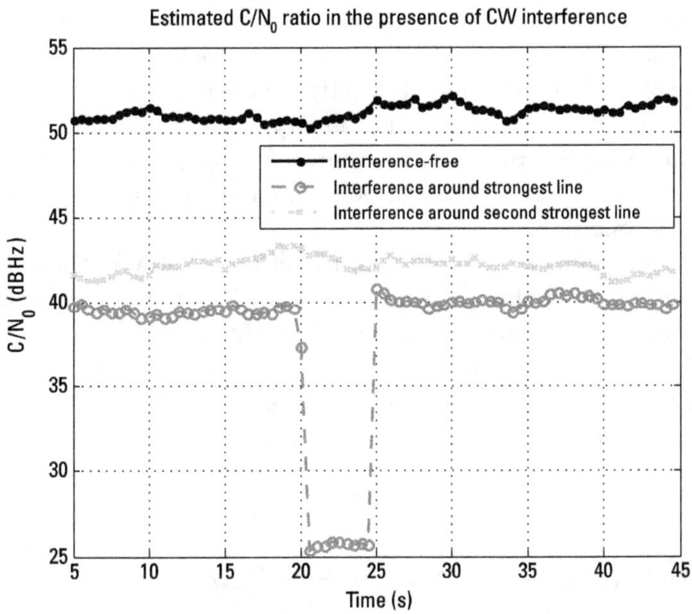

Figure 5.8 Example of C/N_0 degradation due to a CW interference: The interference spectral line passes through a 200-Hz bandwidth around the code line between 20s and 25s.

the second test case considers the CW signal sweeping around the second strongest code line: now the C/N_0 presents a nearly constant degradation of 9 dB, independently from the actual spectral location of the interfering tone.

Two facts can be pointed out:

1. Because of the C/A code signal structure, CWs might be very harmful sources of interference.
2. However, such a harmfulness is strictly related with the relative position between the CW carrier frequency and the strongest spectral component of the code.

5.7 Interference Detection via Pseudorange Monitoring

The quality of the correlator outputs, which determine the level of the C/N_0 measurement, depends on the accuracy of synchronization between SIS and local signal replicas. This synchronization accuracy is also directly responsible for the accuracy of the satellite-to-receiver pseudorange measurements.

As an evident consequence, any impairment affecting the signal tracking loops, including interference, also affects the pseudorange estimates. Thus, as in the case of C/N_0 monitoring, observation of the trends of the pseudorange measurements may bring information about their reliability and the possible presence of a nuisance that impairs accurate SIS reception and pseudorange estimation. Such a nuisance could be an interference, but also any hostile propagation condition.

Pseudorange monitoring to detect interference attacks becomes really effective in the case of dual-frequency receivers, in which simple consistency control of the pseudorange measurements obtained for the same satellite on the two frequencies is able to detect frequency-dependent anomalies. Nonetheless, since frequency-dependent anomalies other than interference may happen (e.g., ionospheric effects), the identification of the anomaly source by pseudorange consistency control is not straightforward.

5.8 Interference Detection via PVT Solution Observation

Interfering signals might induce errors on the estimated PVT, whose accuracy and reliability are the ultimate goals of the GNSS receiver. Like C/N_0 estimates and pseudorange measurements, the number of factors that might affect the PVT accuracy are various (NLOS or blocked SIS propagation, presence of multipath, unexpected dynamics, interference, poor satellite geometry, and so on); therefore, it cannot be used as a stand-alone metric to detect the presence of the interference. Nonetheless, in a controlled environment, measurement of the PVT solution accuracy is the ultimate metric that assesses the effect of a signal degradation in the complete receiving chain. For example, in a controlled condition where interference is the only element that can threaten a signal (e.g., a lab environment or an interference monitoring station), the PVT accuracy measurement can be used in these way:

- To assess the impact of certain families of interferers.
- To test the performance of specific mitigation techniques.
- To cooperate at identifying the presence of harmful interference and activating the proper mitigation technique (specifically in the case of an interference monitoring station) by validating the warnings raised by detection procedures implemented in the early stages of the receiving chain.

An example of position error monitoring to assess the impact of interference is reported in [36].

5.9 Conclusions

Interference detection on GNSS receivers is a complex field, in which contexts, objectives, and techniques are not always the same. First, interference detection (and, even more, mitigation) can be realistically achieved only if a number of different techniques are put in place, designing an "interference alert" mechanism able to integrate the input detection information coming from different detection methods and able to make a unified and confirmed decision about the presence (and intended harmfulness) of what has been discovered. As a consequence of this decision, the proper mitigation mechanism or interference alert will be activated.

Of course, the sensitivity and reliability of such an "interference alert" mechanism is a direct function of the complexity the receiver can allocate. Thus, second, the target application segment of the receiver is a fundamental aspect; for example, it is expected that a mass-market receiver is able to allocate a minimum set (if any) of interference monitoring techniques, tailored to the most common and most easy-to-handle interference types. On the other hand, professional, safety-of-life, and military receivers are expected to allocate increasing capabilities to interference monitoring. On top of this, high-end receivers designed to become monitoring stations of the quality of the GNSS signals must be equipped with the strongest and most sensitive interference detection capabilities and also be able to raise and transmit timely alerts as soon as interference has been detected.

In the end, and as usual, the target application/market segment of a receiver is the driver for the selection of the set of interference detection/mitigation techniques that are implemented on board.

References

[1] Hegarty, C. J., Bobyn, D., Grabowski, J., and Van Dierendonck, A. J., "An Overview of the Effects of Out-of-Band Interference on GNSS Receivers," Proceedings of the 24th International Technical Meeting of The Satellite Division of the Institute of Navigation (ION GNSS 2011), Portland, OR, September 2011, pp. 1941–1956.

[2] Motella, B., et al., "Method for Assessing the Interference Impact on GNSS Receivers," *IEEE Trans. on Aerospace and Electronic Systems*, Vol. 47, 2011, pp. 1416–1432.

[3] Grant, A., et al., "GPS Jamming and the Impact on Maritime Navigation," *J. of Navigation*, No. 62, 2009, pp. 173–187.

[4] Motella, B., M. Pini, and F. Dovis, "Investigation on the Effect of Strong Out-of-Band Signals on Global Navigation Satellite Systems Receivers," *GPS Solutions*, Vol. 12, No. 2, March 2008, pp. 77–86, doi:10.1007/s10291-007-0085-5

[5] De Lorenzo, D. S., et al., "Testing of Adaptive Beamsteering for Interference Rejection in GNSS Receivers" *Proc. ENC 2007*, Geneva, Switzerland, pp. 1277–1287.

[6] Gupta, I. J., et al., "Non-Planar Adaptive Antenna Arrays for GPS Receivers," *IEEE Antennas and Propagation Magazine*, Vol. 52, No. 5, October 2010, pp. 35–51.

[7] Bastide, F., et al., "Automatic Gain Control (AGC) as an Interference Assessment Tool," *Proc. ION GPS 2003*, Portland, OR, 2003, pp. 2042–2053.

[8] Ward, P. W., "Simple Techniques for RFI Situational Awareness and Characterization in GNSS Receivers," *Proc. 2008 National Technical Meeting of the Institute of Navigation*, San Diego, CA, January 2008, pp. 154–163.

[9] Cho, K. M., "Optimum Gain Control for A/D Conversion Using Digitized I/Q Data in Quadrature Sampling," *IEEE Trans. on Aerospace and Electronic Systems*, Vol.27, No.1, January 1991, pp. 178–181. doi:10.1109/7.68164

[10] Lotz, T., Adaptive Analog-to-Digital Conversion and Pre-Correlation Interference Mitigation Techniques in a GNSS Receiver, Master's thesis, Technical University of Kaiserslautern, 2008.

[11] Izos, O., et al., "Assessment of GPS L1/Galileo E1 Interference Monitoring System for the Airport Environment," *Proc. ION GNSS 2011 Conf.*, Portland, OR, September 19–23, 2011, pp. 1920–1930.

[12] Balaei, A., A. Dempster, and B. Motella, "GPS Interference Detected in Sydney-Australia," *Proc. Int. Global Navigation Satellite Systems, IGNSS Symp. 2007*, Sydney, Australia, December 4–6, 2007.

[13] Grabowsky, J., and C. Hegarty, "Characterization of L5 Receiver Performance Using Digital Pulse Blanking," *Proc. Institute of Navigation GPS meeting (ION GPS)*, Portland, OR, September 2002, pp. 1630–1635.

[14] Marti, L, and F. van Graas, "Interference Detection by Means of the Software Defined Radio," *Proc. ION GNSS 17th Int. Meeting of the Satellite Division*, Long Beach, CA, September 2004, pp. 99–109.

[15] Motella, B., and L. Lo Presti, "Methods of Goodness of Fit for GNSS Interference Detection," *IEEE Trans. on Aerospace and Electronic Systems, IEEE Transactions on Aerospace and Electronic Systems,* Vol. 50, No. 3, July 2014, pp. 1690–1700.

[16] Motella, B., M. Pini, and L. Lo Presti, "GNSS Interference Detector Based on Chi-Square Goodness-of-fit Test," presented at 6th ESA Workshop on Satellite Navigation Technologies (NAVITEC 2012), December 2012.

[17] Pestman, W. R., *Mathematical Statistics*, 2nd ed., Berlin: deGruyter, 2009.

[18] Motella, B., and L. Lo Presti, "Pulsed Signal Interference Monitoring in GNSS Application," *Proc. ENC GNSS 2006 Conf.*, 2006, Manchester, UK.

[19] Savasta, S., "GNSS Localization Techniques in Interfered Environments," Ph.D. Dissertation, Politecnico di Torino, January 2010.

[20] Ouyang, X., and M. G. Amin, "Short-Time Fourier Transform Receiver for Nonstationary Interference Excision in Direct Sequence Spread Spectrum Communications," *IEEE Trans. on Signal Processing*, Vol. 49, No. 4, April 2001, pp. 851–863. doi:10.1109/78.912929

[21] Lach, S., R., M. G. Amin, and A. R. Lindsey, "Broadband Interference Excision for Software-Radio Spread-Spectrum Communications Using Time-Frequency Distribution Synthesis," *IEEE J. on Selected Areas in Communications*, Vol. 17, No. 4, April 1999, pp. 704–714. doi:10.1109/49.761046

[22] Kay, S., *Modern Spectral Estimation: Theory and Application*, Upper Saddle River, NJ: Prentice Hall, 1988.

[23] Tani, A., and R. Fantacci, "Performance Evaluation of a Precorrelation Interference Detection Algorithm for the GNSS Based on Nonparametrical Spectral Estimation," *IEEE Systems J.*, Vol. 2, No. 1, March 2008, pp. 20–26. doi:10.1109/JSYST.2007.914772

[24] Balaei, A. T., and A. G. Dempster, "A Statistical Inference Technique for GPS Interference Detection," IEEE Transaction on Aerospace and Electronic systems, Vol. 45, No. 3, July 2009.

[25] Zhang, J., and E. Lohan, "Effect and Mitigation of Narrowband Interference on Galileo E1 Signal Acquisition and Tracking Accuracy," *Proc. 2011 Int. Conf. on Localization and GNSS (ICL-GNSS)*, June 29–30, 2011, pp. 36–41. doi:10.1109/ICL-GNSS.2011.5955262

[26] Cohen, L., *Time-Frequency Analysis*, Upper Saddle River, NJ: Prentice-Hall, 1995.

[27] Borio, D., et al., "Time-Frequency Excision for GNSS Applications," *IEEE Systems J.*, Vol. 2, No. 1, March 2008, pp. 27–37. doi:10.1109/JSYST.2007.914914

[28] Paonni, M., et al., "Wavelets and Notch Filtering. Innovative Techniques for Mitigating RF Interference," *Inside GNSS*, January–February 2011, pp. 54–62.

[29] Musumeci, L., and F. Dovis, "Use of the Wavelet Transform for Interference Detection and Mitigation in Global Navigation Satellite Systems," *Int. J. of Navigation and Observation*, Vol. 2014, 2014. doi:10.1155/2014/262186.

[30] Bastide, F., E. Chatre, and C. Macabiau, "GPS Interference Detection and Identification Using Multicorrelator Receivers," *Proc. 14th Int. Technical Meeting of the Satellite Division of the Institute of Navigation (ION GPS 2001)*, Salt Lake City, UT, September 2001, pp. 872–881.

[31] Linty, N., et al., "Dispositivo di Rilevazione di un Segnale Interferente per un Sistema Globale di Navigazione Satellitare," European patent pending, May 2013.

[32] Manolakis, D., V. Ingle, and S. Kogon, *Statistical and Adaptive Signal Processing*, New York: McGraw Hill, 2005.

[33] Fantino, M., A. Molino, and M. Nicola, "N-GENE GNSS Receiver: Benefits of Software Radio in Navigation," *Proc. European Navigation Conference (ENC 2009)*, Napoli, Italy, May 3–6, 2009.

[34] Groves, P. D., "GPS Signal to Noise Measurement in Weak Signal and High Interference Environments," *Proc. 18th Int. Technical Meeting of the Satellite Division of the Institute of Navigation (ION GNSS 2005)*, Long Beach, CA, September 2005, pp. 643–658.

[35] Falletti, E., M. Pini, and L. Lo Presti, "Low Complexity Carrier-to-Noise Ratio Estimators for GNSS Digital Receivers," *IEEE Trans. on Aerospace and Electronic Systems*, Vol. 47, No. 1, January 2011, pp. 420–437. doi:10.1109/TAES.2011.5705684

[36] Balaei, A. T., et al., "Mutual Effects of Satellite Quality and Satellite Geometry on Positioning Quality," *Proc. ION GNSS 2007 Conf.*, Fort Worth, TX, September 2007.

6

Classical Digital Signal Processing Countermeasures to Interference in GNSS

Luciano Musumeci and Fabio Dovis

Researchers have proposed a large number of digital signal processing techniques to deal with RFI in the GNSS bandwidths. From a general standpoint, such techniques can be classified according to the domain in which the interference mitigation process is implemented. Thus, such techniques can be grouped as follows:

- *Frequency-domain techniques*, in which interference suppression is performed in the frequency domain looking at the characteristics of the spectrum of the interfered GNSS signal that is received;
- *Time-domain techniques*, which operate either by modifying some receiver parameter depending on the characteristics of the received signal in order to mitigate the impact of the interference on the following stages, or by "gating" the signal itself in order to cut off portions of the signal that are believed to be affected by interference;
- *Time-space domain techniques*, which are based on the principle of spatial filtering, thus introducing attenuation in the direction of

arrival of the interfering signals. They typically require a complex hardware configuration because, in general, they are exploiting antenna arrays.

6.1 Frequency-Domain Techniques

The approach to interference mitigation in the frequency domain is quite obvious. Any mitigation technique should be able to filter out the harmonic components of the interfering signal, preserving as much as possible the spectrum of the original GNSS signal. This approach is effective when the interfering signal occupies a limited portion of the frequency spectrum, that is, when it can be classified as narrowband interference (NBI) or continuous-wave interference (CWI).

Nevertheless, an interfering signal might change its spectral characteristic in time, thus requiring flexibility on the part of the mitigation unit to adapt to the actual interfering scenario. This may be the case with low-cost commercial jammers that aim to disturb a wider portion of the spectrum, frequency modulating a narrowband signal in order to span a larger frequency interval over time.

For such a reason in the following sections two common techniques are described, frequency-domain adaptive filtering (FDAF) and notch filtering, with a focus on their capability to adapt to changes in the instantaneous spectrum of the interfering signals.

Frequency-domain techniques are, in general, weakly effective in the case of pulsed interference, because the presence of the interfering signal for a limited time is often lost in the phase of spectral estimation, which is performed by observing the signal in a limited time window and averaging the signal characteristics.

6.1.1 Frequency-Domain Adaptive Filtering

The FDAF is an interference detection and mitigation algorithm based on the estimation of the spectral features of the incoming signal samples at the ADC output. The spectral estimation is obtained by means of the fast Fourier transform (FFT) on a predefined number of samples N (observation window). Figure 6.1 shows the FDAF functional scheme.

The amplitude of each point of the signal's Fourier representation is compared to a threshold determined according to the expected received signal power and the estimate of the noise floor in an interference-free environment.

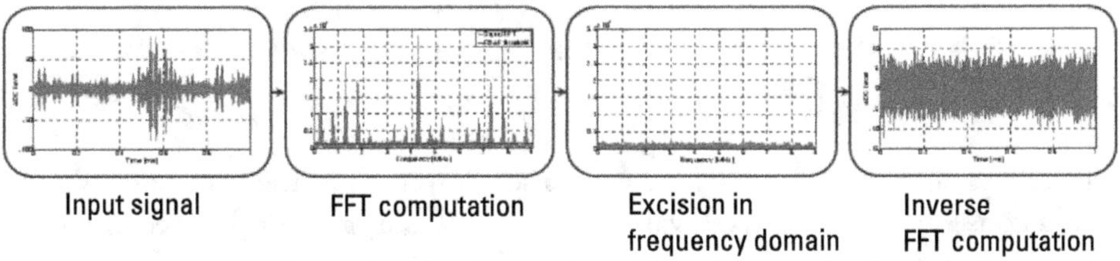

Figure 6.1 FDAF functional scheme.

Because the entire GNSS signal is below the thermal noise floor, the FFT representation should be, ideally, almost flat. If certain points of the Fourier transform of the incoming signal exceed this threshold, they are considered corrupted by interference and set to zero. Finally, the inverse FFT of the manipulated incoming signal is performed so as to obtain the signal back in the time domain.

The effectiveness of this technique is strictly related to the ability to estimate the spectrum of a random process using a discrete Fourier transform (DFT), and thus the typical criteria for the choice of the DFT parameters have to be used. A complete description of the theory of spectral estimation based on DFT can be found in [1]. The main parameters impacting the performance of such a technique operating in the frequency domain are represented by the employed sampling frequency f_s and the number of points N_D used to compute the Fourier transform. In fact, the spectrum of a sampled random process can be estimated on the basis of a set of N samples of the incoming signal $x[0] \ldots x[N-1]$. According to *Bartlett's procedure*, such a data stream is split into N_{seg} segments of N_D samples each. The sample spectrum $S_x^{(p)}(f)$ of each segment $x^{(p)}[n], p = 1, \ldots, N_{seg}$ is computed, and the spectrum is estimated as the average:

$$S_x(f) = \frac{1}{N_{seg}} \sum_{p=0}^{N_{seg}-1} S_x^{(p)}(f) \tag{6.1}$$

The resolution Δf of the spectral estimation is then determined by $\Delta f = f_s / N_D$ (Hz). If the FFT resolution, Δf, is too large compared to the spectral characteristics of the interfering signal, detection of the interference frequency components may not be successful. Furthermore, the variance of the power spectral estimator calculated with Bartlett's method is inversely proportional to the number of segments:

$$\operatorname{var}\{S_x(f)\} \simeq \frac{S_x^2(f)}{N_{\text{seg}}} \tag{6.2}$$

It follows that the quality of the estimation improves as the inverse of N_{seg}. As a consequence, for a fixed value $N = N_D \cdot N_{\text{seg}}$, a trade-off between high-frequency resolution (N_D as large as possible) and low variance (N_{seg} as large as possible) must be found.

The average periodogram is a biased estimate of the power spectrum due to the presence of a *rectangular window* of length N_D, implicit in the selection of a limited number of samples.

Different types of windows can be used to weight the samples in order to mitigate edge effects and limit the bias. However, the bias cannot be totally removed and it has to be taken into account [1].

For the sake of providing an example, Figure 6.2 shows the spectral estimation by means of the Bartlett periodogram of a received GPS C/A code affected by in-band CWI close to the intermediate frequency of 4 MHz. Note how, given a fixed number of samples N of the incoming signal, the choice of N_D and N_{seg} is the result of a trade-off between bias and variance of the estimation. It is clear how such a choice has as a consequence on the capability of detecting the presence on the CWI. Choosing $N_{\text{seg}} = 1$ provides the best resolution, but also a noisy estimation of the spectrum that might mask the

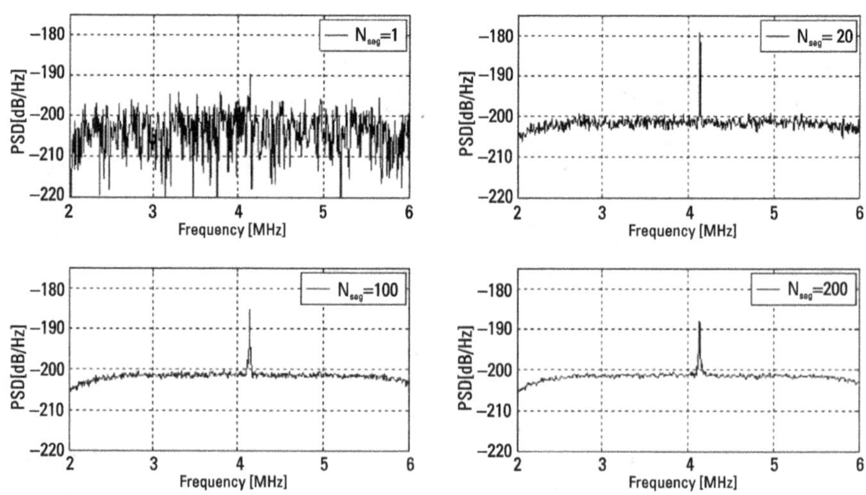

Figure 6.2 Bartlett-based spectral estimation as a function of N_{seg}.

presence of the interference. Increasing N_{seg} allows for a better identification of the CWI, until the biasing effect of the window dominates and it could lead the FDAF to excise too many values of the spectrum.

It is then clear that the optimal choice of the parameters would require knowledge of the nature of the interfering signal. For NBI or CWI the frequency resolution might be relevant for a good detection performance. Either a rough estimation of the spectrum with a limited number of N_{seg} or a value of N_{seg} that is too large, enhancing the biasing effect, could result in several false alarms and to the filtering of too many frequency components.

However, increasing the FFT frequency resolution by means of a large number of FFT points N_D also results in a higher computational burden for the FFT. Thus, the value N_D also has to be determined by trading off the interference detection capability and the implementation complexity.

In time-varying scenarios, the performance and the capability of the technique to be "adaptive" with respect to the nonstationary interference are also traded off with the rate at which the estimation of the spectrum is updated, that is, based on the total length of the observation window N and on the time required for the FFT computations.

The ADC stage and the automatic gain control (AGC) block also have an impact on FDAF performance. The quantization process of the IF signal causes small signal-to-noise (SNR) ratio degradations that depend on the factor $k = L/\sigma$ where L is the maximum quantization level and σ is the signal's standard deviation at the input of the quantizer. Thus, depending on the number of the adopted quantization bit, the optimal ratio k that provides the minimum SNR degradation has to be identified. The presence of an interfering signal would also negatively impact the AGC gain tuning process, which assumes that only Gaussian noise is present. This is mainly due to the fact that, in the presence of an interfering signal, the ADC output, which drives the AGC gain, is no longer Gaussian distributed. To overcome this problem in [2], a scheme where the AGC gain is driven by the FDAF algorithm output is proposed.

Even if the FDAF is demonstrated to be effective for NBI and CWI removal, in [2] it is applied against pulsed interference. In such a case the goal is to remove the spikes that are generated in the spectrum of the received signal by the pulsed interference. The authors show that the FDAF algorithm's performance for pulsed interference suppression is better with respect the performance achieved by the blanker acting in the time domain. However, consistent gain (over 2.5 dB with respect to the pulse blanking) is achieved only when a high number of points for the FFT computation are employed, thus leading to a very complex implementation for the algorithm.

6.1.2 Notch Filtering

Notch filtering has been proved to be an efficient mitigation algorithm for a family of interfering signals, pure sinusoids, which appears as a spike in the spectral domain. This kind of interfering signal, usually generated by TV transmitters, VOR and ILS stations, are spurious signals caused by power amplifiers working in nonlinearity regions or by oscillators present in many electronics devices.

CWI can be considered harmful for the acquisition and tracking stage operation of a GNSS receiver, due to its heavy impact on the receiver's correlation process. In [3] the authors analyze the impact of CWI on the acquisition stage of a GNSS receiver providing a detailed theoretical derivation of the detection probability degradation; in [4] the authors provide two methods, the interference error envelope (IEE) and the interference running average (IRA), for assessing the impact of CWI at the tracking level. Notch filters are usually characterized by a passband frequency response with a very narrow portion of rejection spectrum correspondence of the CWI carrier frequency, thus providing attenuation of the interfering signal and preserving as much as possible the useful GNSS signal spectral components. An example of notch filter frequency response is shown in Figure 6.3.

The most common implementation of notch filters is by means of infinite impulse response (IIR) digital filters. A causal IIR filter can be written in terms of a general difference equation where the output signal at a given instant is the linear combination of samples of the input and output signal at previous time instant:

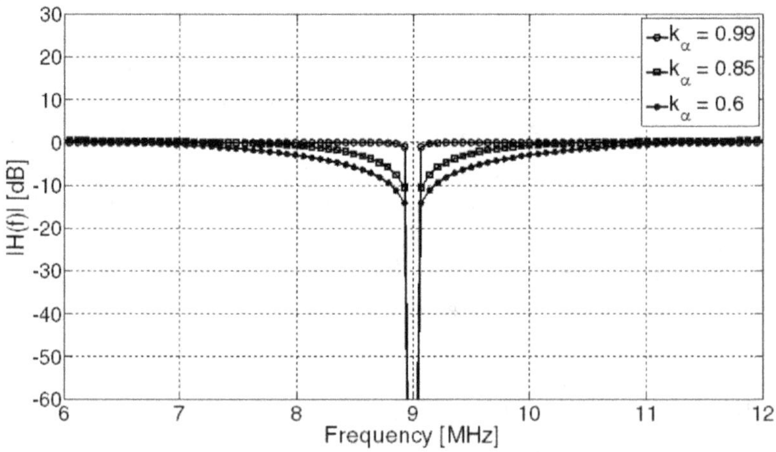

Figure 6.3 Notch filter frequency response.

$$y[n] = -\sum_{m=1}^{N} a_m y[n-m] + \sum_{m=0}^{m} b_m x[n-m] \tag{6.3}$$

Thus, in the Z-transform domain, the IIR filter response becomes

$$H(z) = \frac{b_0 + b_1 z^{-1} + \ \ldots\ + b_m z^{-m}}{1 + a_1 z^{-1} + \ \ldots\ + a_m z^{-m}} \tag{6.4}$$

A real CWI presents two spectra lines corresponding to the frequency f_i and $-f_i$. Thus, in [5] the authors propose a two-pole notch filter for CWI mitigation, the transfer function of which is

$$H(z) = \frac{1 - 2\Re\{z_0\}z^{-1} + |z_0|^2 z^{-2}}{1 - 2k_\alpha \Re\{z_0\}z^{-1} + k_\alpha^2 |z_0|^2 z^{-2}} \tag{6.5}$$

where z_0 is placed in correspondence with the interfering frequency $z_0 = \beta\exp\{j2\pi f_i\}$. The parameter $0 < k_\alpha < 1$, known as the pole contraction factor, determines the width of the notch filter. The closer k_α is to unity, the narrower the notch filter, which in turns means a reduction of the distortion on the useful GNSS signal. However, k_α cannot be chosen arbitrarily close to unity for stability reasons and thus a compromise has to be found.

An approximate expression that relates the factor k_α with the 3-dB attenuated bandwidth of the notch filter B_{3db_Hz} can be empirically set as

$$B_{3dB\_Hz} \approx \frac{(1 - k_\alpha)\pi}{10} \cdot f_s \tag{6.6}$$

In the presence of multiple tones, a multipole notch filter, based on the use of several two-pole notch filters in cascade, can be used. In such a case the first two-pole notch filter in the chain mitigates the most powerful disturbing signal, whereas the others remove the residual sinusoids with progressively decreasing power.

6.1.3 Adaptive Notch Filter

Because the interference carrier frequency is an unknown parameter, the authors in [5] also propose an integration of the two-pole notch filter with an adaptive unit, which is in charge of the CWI carrier frequency estimation. A

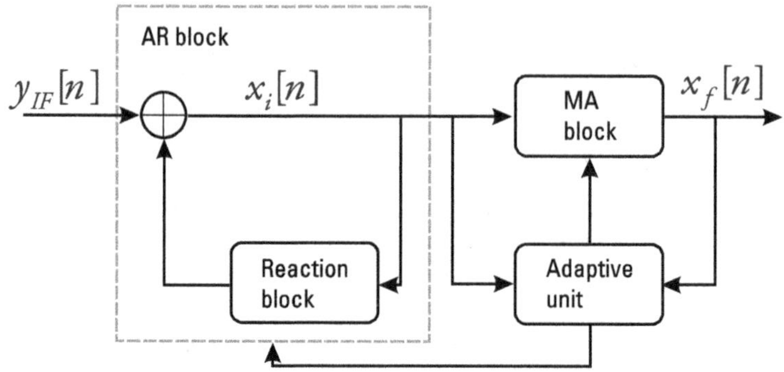

Figure 6.4 Notch filter structure.

basic structure for the two-pole notch filter coupled with an adaptive unit is reported in Figure 6.4.

The numerator of the filter transfer function represented in (6.4) is defined as the moving average (MA) block of the two-pole notch filter, whereas the denominator in (6.4) represents the autoregressive (AR) block, which is introduced in order to compensate for the effect of the MA part. In this structure, the detection algorithm for the determination of the CWI frequency component perturbing the received GNSS signal is based on the removal of the constraint on the location of the filter zeros in the complex plane, and their amplitude is adjusted by an adaptive unit. Through this algorithm, the notch filter is able to detect the presence of the interfering signal and to decide whether to use its filtered output or the unmodified input signal.

The adaptive algorithm, proposed in [5], is based on an iterative normalized least mean square (LMS), which minimizes the cost function $f_C[n] = E\{|x_f[n]|^2\}$ where $x_f[n]$ is the output of the filter. The minimization is performed with respect to the complex parameter z_0, using the iterative rule

$$z_0[n+1] = z_0[n] - \mu[n] \cdot g\left(f_C[n]\right) \tag{6.7}$$

where

$$g\left(f_C[n]\right) = 4x_f[n]\left(z_0[n]x_i[n-2] - x_i[n-1]\right) \tag{6.8}$$

is the stochastic gradient of the cost function $f_C[n]$, and $\mu[n] = \dfrac{\delta}{\mathbb{E}_{x_i[n]}}$ is the algorithm step with $\mathbb{E}_{x_i[n]}$ being an estimate of $\mathbb{E}\{|x_i[n]|^2\}$, which is in turn the power of the AR block output $x_i[n]$. The parameter δ is the unnormalized LMS algorithm step that controls the convergence properties of the algorithm.

The amplitude of z_0 is extremely dependent on the interfering power. In fact, as the interfering power decreases, the minimum of the cost function $f_C[n]$ is no longer achieved by removing only the interference but also by attenuating a part of the noise and GNSS signal components. Attenuation of the GNSS signal components can be avoided by monitoring the value of the amplitude of the zero z_0. If it passes a fixed threshold, it means that the adaptive notch filter is tracking a CWI signal and thus its output has to be used by the GNSS receiver. Otherwise, the input unfiltered signal has to be employed.

The detection threshold can be fixed by choosing an interference-to-noise ratio $J/N = \gamma$ that can be considered harmful for the GNSS receiver. The threshold T is determined as the value of the notch filter amplitude of the zero when the ratio $J/N = \gamma$. Then a simple test verifies the condition $|\hat{z}_0| > T$, where $|\hat{z}_0|$ is the estimation of the mean of z_0 to enable the notch filter. The scheme of the adaptive notch filter coupled with the interference detection unit is shown in Figure 6.5.

The presence of the adaptive block also makes the entire notch filter suitable for suppressing the harmful interference produce by jammers. Such

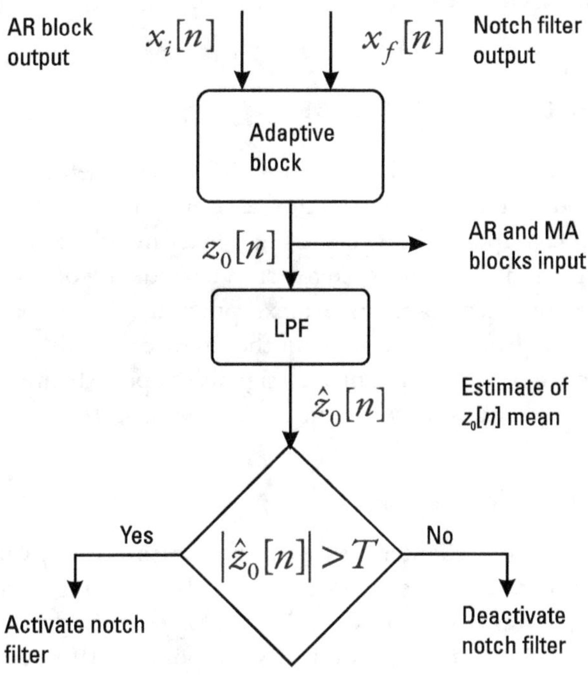

Figure 6.5 Adaptive notch filter detection algorithm.

devices, available on the web for a few dollars, transmit strong chirp signals sweeping several megahertz in a few microseconds, thus appearing in the spectrum as wideband interference (WBI). More details on the use of such two-pole notch filters for jamming suppression can be found in [6].

Several variations of this notch filter (adaptive or not) have been proposed. Reference [7] describes how the position of z_0 with respect to the unit circle affects the distortion of the signal at the notch filter output. Here a different adaptive algorithm, consisting of forcing the zero of the filter to move on the unit circle, is proposed. Furthermore, to improve the convergence speed of the adaptive algorithm, a runtime change of the pole contraction factor k_α and of the LMS step δ is performed. In the absence of interference, the notch width is wide and the LMS step is large. When the interference appears, the notch becomes narrower, the convergence step smaller, and the zero is forced to move on the unit circle.

Although the notch filter represents an effective countermeasure when dealing with CWI, it does not represent the best solution for coping with multiple interfering signals jamming all of the GNSS received signal bandwidth. In this case, implementation of notch filters for suppressing multiple NBI spread all over the GNSS useful signal spectrum would become extremely complicated, as mentioned in [8].

6.2 Time-Domain Techniques

Observation of the signal in the time domain is often useful for interference detection purposes, but is not always the best domain for applying mitigation techniques. In fact, most of the interfering signals are mixed to the incoming GNSS signals, and it is not possible to act independently on the interference and on GNSS signal components. An exception are the pulsed interference signals, which, in general, are limited in the time domain, but they do affect the whole frequency spectrum. For such signals the pulse blanking technique has been proposed and is presented in the following section.

6.2.1 Pulse Blanking Technique

The most common pulsed interference countermeasure, already implemented in modern GNSS receivers, is represented by pulse blanking circuitry. Such a simple technique was first proposed using analog technology as explained in [9]; a fully digital implementation was first proposed in [10]. A block scheme

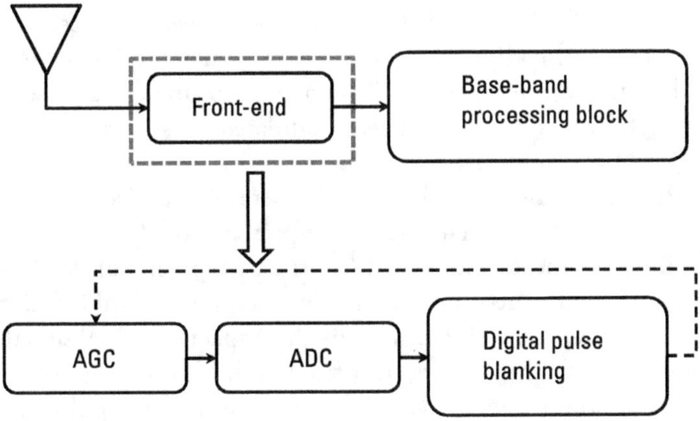

Figure 6.6 Digital pulse blanking implementation.

of the digital pulse blanking implementation within the digital GNSS receiver front-end is shown in Figure 6.6.

This type of digital circuitry provides pulsed interference excision by means of a thresholding operation on the samples at the ADC output. Basically, each sample is compared to a digital threshold level, which is set according to an estimation of the thermal noise power only, and it is blanked whenever the threshold is exceeded.

The principle of the digital pulse blanking is simple. It relies on the fact that pulses are short and have very large amplitude compared to the noise level. For such a reason, the implementation of pulse blanking circuitry requires the presence of an ADC quantizing the incoming signal over a large number of bits. In this way, the AGC can be tuned in order to map the received signal level exploiting a limited number of bits (e.g., 2 or 3), leaving the higher bits for pulse detection purposes. Otherwise, the AGC, tuned in order to exploit the full ADC scale, would considerably suppress the useful GNSS signal during the on active state of the pulse, thus masking the presence of the pulse itself to the blanking circuitry.

The detection threshold can be chosen as a compromise between the ability to detect pulses and the C/N_0 degradation in the absence of pulses.

The typical use of pulse blankers is in GNSS receivers designed to operate in aviation scenarios. In fact, many aeronautical radio navigation systems (ARNS) based on strong pulsed signal transmission from ground beacons, such as DME or TACAN introduced earlier in Chapter 2, share the same frequency bandwidths of GPS L5 and Galileo E5 signals [11]. In such a scenario, the interference affecting the on-board GNSS receiver is represented by the composite strong pulsed signals transmitted from all of the ARNS ground stations visible to the receiver. Figure 6.7 provides an example of a pulse blanking operation performed on a normalized modulated pulse, transmitted by the DME/TACAN ground beacon.

The drawbacks of an ideal blanking operation applied on such a pulse is represented by the fact that due to the presence of the modulation over the pulse duration and to the typical Gaussian shape, not all of the samples belonging to the pulse are suppressed. Moreover, pulse blanking circuit performance can be negatively influenced by the impact of pulsed signals on the active components within the receiver front-end. Very strong pulses or very strong received power due to the combination of multiple pulses can cause the saturation of the active components in the GNSS receivers (e.g., amplifiers), which may require a recovery time to go back to a normal state when the interference ends. Reference [9] mentions that for a particular commercial receiver, an interference pulse signal with a peak power of 15 dB above the thermal noise is sufficient to saturate the last amplification stage within the receiver front-end. Under this interference environment condition, pulse blanking may perform signal suppression even during the off state of the pulse for a time period equal to the recovery time needed by the amplifiers to resume normal operation. For a commercial receiver, typical recovery times for amplification stages is about 40 ns/dB of input level beyond the saturation point [9].

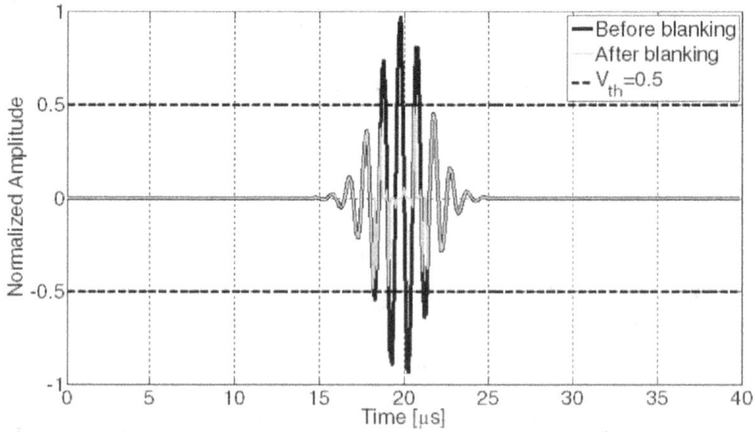

Figure 6.7 DME/TACAN pulse before and after blanking.

In general, the impact of pulsed interference signals on receiver front-end components might be different depending, in particular, on the pulse peak power level and on the pulse duration. Furthermore, the design of the AGC/ADC block is also critical to pulse blanking performance. The AGC, needed when multibit quantization is implemented in the digital part of the receiver front-end, has to be carefully designed. A slow AGC sets the ADC input levels averaging the input signal power over a large time during which if too many pulses oscillations are present, the input dynamics of the ADC are not properly set [9].

The fact that the blanked samples should not be used for AGC tuning in order to avoid ADC overloading must also be considered. For these several reasons, a blanking operation might not be efficient, because a high percentage of the received GNSS signal may be blanked even during the off state of the pulse interference, thus considerably increasing the distortion of the received signal [10]. Figure 6.8 provides a comparison of the effects of ideal pulse blanking and nonideal blanking on a DME/TACAN single pulse.

The nonideal blanking, representing the effect of the interference on the front-end components, leads not only to a delay in the detection of the initial part of the pulses, but also to a delay in the pulse blanking deactivation time, thus leading to nonideal suppression of the interference and to a greater degradation of the useful GNSS signal. However, digital pulse blanking

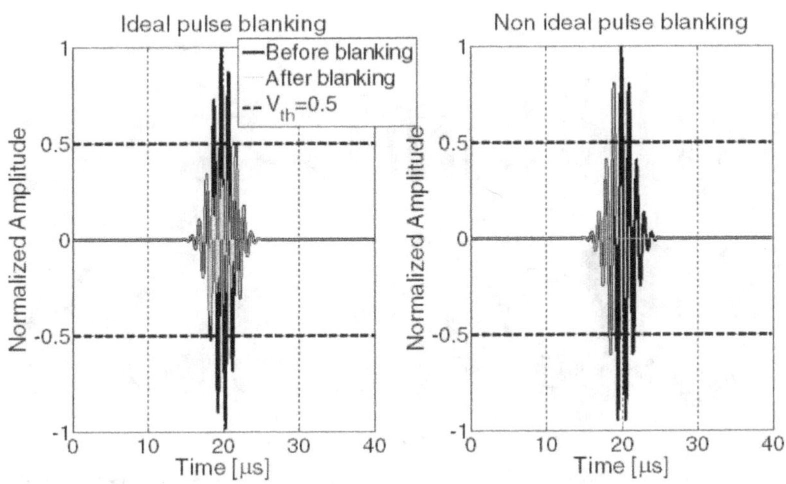

Figure 6.8 Ideal versus nonideal pulse blanking.

represents a low-cost, low–complexity, highly effective solution for most of the pulsed interference environment experienced at low-altitude scenarios, such as during landing procedures for avionics equipment. In such scenarios, due to the low altitude, the GNSS receiver has only a few ARNS beacons in LOS (two or three nearby the airport) and the strong pulsed signals achieving the GNSS airborne antenna do not represent a pulsed interference environment very dense in time, which can then be mitigated by the digital pulse blanking without causing high degradation in the SNR. Figure 6.9 shows the profile of the carrier-to-noise density ratio degradation (top plot) and the blanker duty cycle, defined as the average percentage of suppressed received signal (bottom plot), caused by digital pulse blanking along a simulated landing procedure for a discretized set of altitude values. At low altitude, around 1,060m, the digital pulse blanking suppresses around 10% of the received signal, leading to a degradation of less than 2 dB on the tracked Galileo E5a and GPS L5 signals.

The pulse blanking is limited in performance in scenarios where a large number of pulsed interfering signals are present, thus creating, for the GNSS receiver, a very dense in time pulsed interference [12] that will trigger the pulse blanker for long portions of time. References [13] shows how in some areas located close to large airports digital pulse blanking is suppressing more than the 56% of the received signal, causing a heavy degradation of the SNR that leads, in some cases, to a complete loss of lock of the weakest GNSS received signal [13].

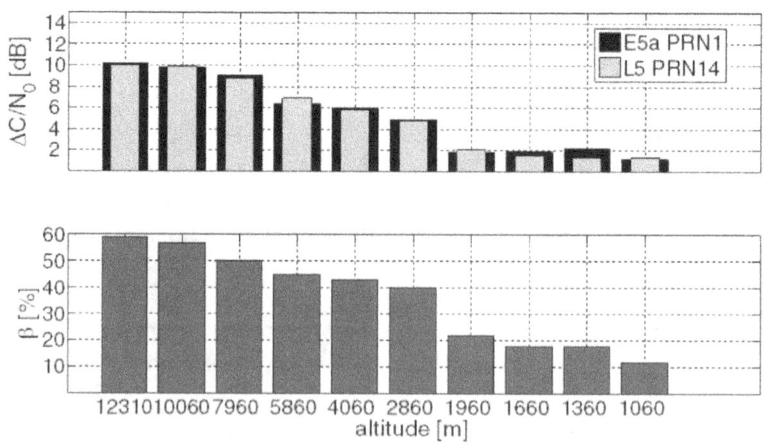

Figure 6.9 Landing procedure simulation: carrier-to-noise density ratio degradation and blanker duty cycle versus aircraft altitude.

6.3 Space-Time Domain Techniques

Space-domain techniques require a highly complex hardware configuration that exploits the antenna array. Two families of space-domain techniques can be identified:

- Null steering techniques, which exploit the use of a controlled radiation pattern array (CRPA);
- Digital beamforming.

The use of CRPA is a very effective technique against continuous interference. This technique nulls out the signal in the direction of the interference and is able to mitigate WBI or NBI. It can be also used against pulsed interference sources like DME; however, in such a case the CRPA technique is often not able to produce effective adaptive interference nulling because of the convergence time required by an antenna control algorithm. As a rule, the CRPA technique is the reference technique in the context of RF/IF analog beamforming. The major advantage of the analog approach is that it can be designed to prevent saturation effects in the RF/IF part of the receiver signal processing and the distortion of A/D conversion process. The utilization of the CRPA technique significantly increases antenna costs but can provide up to 30 dB of interference rejection.

Regrettably, a CRPA is larger than comparable single-element antennas. The disadvantage of the analog CRPA technique is that all satellite signals are processed in a single RF/IF channel and therefore simultaneous interference nulling and producing of multiple beams toward satellites is not feasible.

Digital beamforming is a variation of the CRPA technique in which the beamforming takes place in the digital signal processing part of the receiver. The use of digital beamforming enables individual satellite signals to be processed in separate signal processing channels. As a result, in addition to the "simple" null-steering effect in the direction of arrival of the interfering signal, the digital beamforming in each channel can be optimized to the reception of a particular GNSS signal, for example, by producing additional antenna gain into the satellite direction. This digital approach is much cheaper than traditional analog CRPAs, but it does not prevent front-end saturation from high-power interference. Therefore, a combination of digital beamforming with high dynamic LNAs and application of a multibit ADC with AGC is recommended, resulting in interference mitigation performance of 12 and 28 dB for a 4- and a 16-element array, respectively [14].

A full investigation of all of the possible implementations of beamforming is beyond the scope of this chapter, but in the following subsections two approaches that have been tested on GNSS signals are described:

- Space-time adaptive processing (STAP) techniques;
- Spatial filtering through subspace decomposition.

6.3.1 Space-Time Adaptive Processing Techniques

The authors of [15] describe two STAP techniques that provide simultaneous pulsed and CWI suppression in the spatial domain exploiting a GNSS antenna array receiver. Figure 6.10 shows a typical configuration for a GNSS receiver implementing a STAP algorithm.

The antenna array is composed of M elements followed by an RF front-end that provides amplification and downconversion to an intermediate frequency. The digital signal at the output of each front-end is fed to an adaptive finite impulse response (FIR) filter with K time taps. The signal at the output of each filter is then summed to produce a digital STAP output, which can be written as follows:

$$y[n] = \sum_{m=1}^{M}\sum_{k=1}^{K} w_{mk} x_m [n-k+1] = \mathbf{W}^{\mathrm{T}}\mathbf{X} \tag{6.9}$$

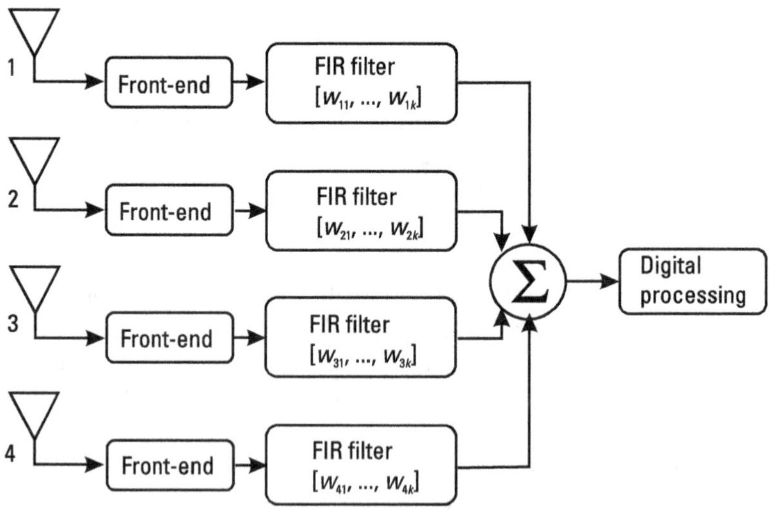

Figure 6.10 GNSS receiver with space-time array processing.

where w_{mk} is the STAP weight at the kth tap of the FIR filter after the mth array element, $x_m[n]$ is the nth sample of the mth array element output, and the STAP input and weight ($MK \times 1$) vectors, **X** and **W** respectively, are defined as follows:

$$\mathbf{X} = \left[x_1[n],\ldots, x_1[n-k+1],\ldots, x_m[n],\ldots,x_m[n-k+1]\right]^T \tag{6.10}$$

$$\mathbf{W} = \left[w_{11},\ldots,w_{1k},\ldots,w_{M1},\ldots,w_{MK}\right]^T \tag{6.11}$$

The interference rejection is operated by the control algorithm depicted in Figure 6.11, which is in charge of updating the weights of each FIR filter in the STAP scheme.
Two weight control algorithm are proposed in [15]:

- The minimum mean square error (MMSE) algorithm updates the STAP weight in order to minimize the mean square difference between the desired reference signal $\mathbf{S_{ref}}$ and the STAP output. In this case the following optimization problem has to be solved:

$$\mathbf{W}_{\text{opt}} = \arg\min_{\mathbf{W}} \mathbb{E}\left\{\left|\mathbf{s}_{\text{ref}} - \mathbf{W}^{\text{T}}\mathbf{X}\right|^2\right\} = \mathbf{R}^{-1}\mathbf{G}_{\text{S}} \tag{6.12}$$

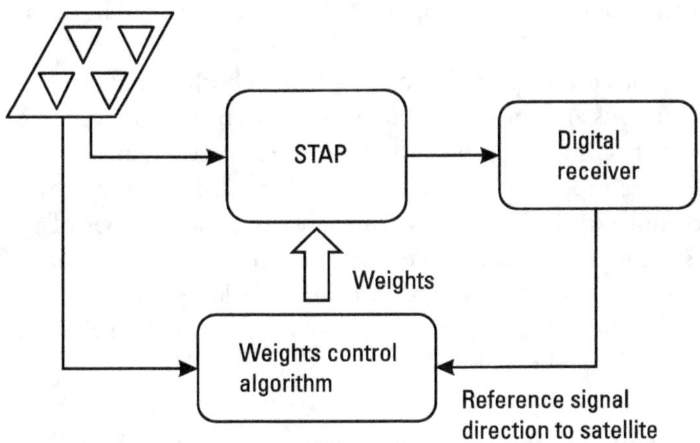

Figure 6.11 STAP weight control algorithm.

where $\mathbf{R} = \mathbb{E}\{\mathbf{X}\,\mathbf{X}^{\mathbf{H}}\}$ is the STAP covariance matrix, and $\mathbf{G_s} = \mathbb{E}\{\mathbf{X}\,\mathbf{s_{ref}}\}$ is the cross-correlation vector between the STAP input and the reference signal.

- The minimum variance distortionless response (MVDR) technique minimizes the STAP output power while preserving a predefined gain at the desired direction. In this case, the optimization problem is formulated as follows:

$$\mathbf{W}_{\text{opt}} = \arg\min_{\mathbf{W}} \mathbf{W}^{\mathbf{H}}\mathbf{R}\mathbf{W} \text{ subject to } \mathbf{W}_{\mathbf{i}}^{\mathbf{T}}\mathbf{A} = \mathbf{1} \text{ and } \mathbf{W}_{\mathbf{j}}^{\mathbf{T}}\Big|_{j\neq i} = 0 \quad (6.13)$$

The vector $\mathbf{W}_i = [w_{1,i}, w_{2,i}, w_{m,i}, w_{M,i}]^T$ is the weight vector at the ith tap that is the central tap in the STAP FIR filter; $\mathbf{A}$ is an $(M \times 1)$ array steering vector in the desired direction and $\mathbf{0}$ is an $(M \times 1)$ vector with all nulls.

The covariance matrix $\mathbf{R}$ and the cross-correlation matrix $\mathbf{G_s}$ are estimated in an iterative way as follows:

$$\hat{\mathbf{R}}[l+1] = \gamma\hat{\mathbf{R}}[l] + \mathbf{X}[l]\mathbf{X}^{\mathbf{H}}[l] \quad (6.14)$$

$$\hat{\mathbf{G}}_{\mathbf{s}}[l+1] = \gamma\hat{\mathbf{G}}_{\mathbf{s}}[l] + \mathbf{X}[l]\mathbf{S}_{\mathbf{ref}}[l] \quad (6.15)$$

where γ is a memory factor that defines to which extent the older estimations are accountable for obtaining a new one; $\mathbf{X}[l]$ is an $(MK \times N)$ matrix that collects N STAP inputs over some adaptation time interval; and $\mathbf{S_{ref}}[l]$ is an $(N \times 1)$ vector containing N samples of the reference signal in the adaptation interval.

In [15] the authors provide a comparison of both STAP methods in mitigating continuous and pulsed interference. It has been shown that, increasing the number of taps employed in the STAP architecture, an improvement of the interference suppression capability is achieved. However, higher computational complexity is required when increasing the number of taps in the FIR filters. Thus a trade-off between performance and complexity is needed.

These two types of digital beamforming belong to the class of adaptive schemes, since they are based on the use of a feedback algorithm that identifies the optimal weight vector $\mathbf{W_{opt}}$ leading to the best CRPA beam capable of rejecting the interference coming from a determined direction and keeping constant the gain in the direction of the useful GNSS signal. As an example of the application to GNSS, in [16] the authors present a real-time, fully software receiver for GPS L1 C/A signal processing that implements digital beamforming based on the MVDR technique and coupled with a CRPA composed by a seven-element antenna array. The implementation of the CRPA is based on the use of the integrated carrier phase (ICP) measurements, usually adopted for code measurement smoothing, obtained from the phase-locked loop. In particular, the ICP differences between different antennas are considered for generating the steering vector, which is fed to the deterministic or adaptive beamforming block.

6.3.2 Subspace Decomposition for Spatial Filtering

Given an antenna array with M sensor elements, it is possible to define the digital signal at the input of the interference mitigation block at epoch k in a matrix form:

$$\mathbf{X}[k] = \mathbf{S}[k] + \mathbf{Z}[k] + \mathbf{N}[k] \tag{6.16}$$

where $\mathbf{X}[k]$, as well as $\mathbf{S}[k]$, $\mathbf{Z}[k]$, and $\mathbf{N}[k]$, are $(M \times N)$ complex matrices containing, respectively, the composite received signal, the useful GNSS received signal component, the interference component and the noise component coming from the M different front-end stages connected to each of the sensors present in the antenna array.

Due to the uncorrelation between useful GNSS signal, interference, and noise components, the spatial covariance matrix of the received signal considering the kth period can be given by

$$\mathbf{R}_{XX}[k] = \mathbb{E}\{[\mathbf{x}[(k-1)N+n]\mathbf{x}^{\mathrm{H}}[(k-1)N+n]\} = \mathbf{R}_{ss}[k] + \mathbf{R}_{zz}[k] + \mathbf{R}_{nn}[k] \tag{6.17}$$

However, because the power of the GNSS signal is completely buried in the noise floor, and it is extremely small with respect to the interference power, the spatial covariance matrix can be approximated as $\mathbf{R}_{xx}[k] \approx \mathbf{R}_{zz}[k] + \mathbf{R}_{nn}[k]$. Thus, the eigen decomposition of the spatial covariance matrix becomes

$$\mathbf{R}_{\mathrm{xx}}[k] \approx \begin{bmatrix} \mathbf{U}_{\mathrm{I}} & \mathbf{U}_{\mathrm{N}} \end{bmatrix} \begin{bmatrix} \mathbf{\Lambda}_{\mathrm{I}} & 0 \\ 0 & 0 \end{bmatrix} \begin{bmatrix} \mathbf{U}_{\mathrm{I}}^{\mathrm{H}} \\ \mathbf{U}_{\mathrm{N}}^{\mathrm{H}} \end{bmatrix} \tag{6.18}$$

where the columns of the unitary matrix $\mathbf{U_I} \in \mathbf{C}^{M \times I}$ span the interference subspace, the columns of the unitary matrix $\mathbf{U_N} \in \mathbf{C}^{M \times (M-I)}$ span the noise subspace, and $\mathbf{\Lambda_I}$ denotes a diagonal matrix that contains the nonzero eigenvalues $\lambda_1, \ldots \lambda_i, \ldots \lambda_I$ with respect to the interference subspace in the noise-free case. For all of the eigenvalues $\lambda_i \gg \sigma_n^2$, a prewhitening matrix to suppress interference in $\mathbf{X}[k]$ can be derived as follows:

$$\mathbf{R}_{\mathbf{XX}}^{-\frac{1}{2}} \approx \frac{1}{\sqrt{\sigma_n^2}} \mathbf{U_N} \mathbf{U_N^H} = \frac{1}{\sqrt{\sigma_n^2}} \mathbf{P_I^{\perp}}[k] \tag{6.19}$$

where $\mathbf{P_I^{\perp}}[k]$ is the projector onto the interference-free subspace for the kth period.

Thus, interference suppression can be achieved by applying the projector matrix to the received digital signal as follows:

$$\tilde{\mathbf{X}}[k] = \mathbf{P}_I^{\perp}[k]\mathbf{X}[k] \tag{6.20}$$

The projector matrix $\mathbf{P_I^{\perp}}[k]$ can be derived from an eigen decomposition of an estimate of the precorrelation spatial covariance matrix of the kth period. The most critical part of such a beamforming technique is contained in the eigen decomposition of $\mathbf{R_{XX}}$, which allows for the determination of the projector matrix $\mathbf{P_I^{\perp}}[k]$. Solving an eigen decomposition problem requires a high computational load, and its software implementation becomes quite critical. In [17] a hybrid implementation based on a software implementation supported by an additional coprocessor is presented.

6.4 Conclusions

This chapter introduced the most common digital signal processing techniques currently being implemented for interference mitigation in GNSS receivers. The overview covered the main strategies devised in the frequency, time, and space domains. The choice of the best technique, in terms of interference removal capability, is dependent on the features of the interfering signal, which in most cases are unknown. For this reason, when implementing an interference mitigation unit, a combination of the different techniques is needed, unless specific information on the interfered environment is available (as in the case of the GNSS receiver operating in the ARNS bandwidth).

The most effective techniques for interference mitigation require a non-negligible increase in the computational complexity, and for such a reason their implementation is nowadays limited to high-end receivers for which fast processing units are available and there are no strict power consumption restrictions. Nevertheless, due to the increased attention to interference phenomena as well as the always increasing computational power of the processors, they are expected be implemented in a growing number of GNSS receivers, even in the mass-market segment.

References

[1] Oppenheim, A. V., and R. Schafer, *Discrete-Time Signal Processing*, Upper Saddle River, NJ: Pearson Education, 2006.

[2] Raimondi, M., et al., "Mitigating Pulsed Interference Using Frequency Domain Adaptive Filtering," *Proc. 19th Int. Technical Meeting of the Satellite Division of the Institute of Navigation (ION GNSS 2006)*, Fort Worth, TX, September 26–29, 2006, pp. 2251–2260.

[3] Borio, D., "GNSS Acquisition in the Presence of Continuous Wave Interference," *IEEE Trans. on Aerospace and Electronic Systems*, Vol. 46, No. 1, 2010, pp. 46–60.

[4] Motella, B., et al., "Method for Assessing the Interference Impact on GNSS Receivers," *IEEE Trans. on Aerospace and Electronic Systems*, Vol. 47, No. 2, 2011, pp. 1416–1432.

[5] Borio, D., L. Camoriano, and L. Lo Presti, "Two Pole and Multi Pole Notch Filters: A Computationally Effective Solution for GNSS Interference Detection and Mitigation," *IEEE Systems J.*, Vol. 2, No. 1, 2008, pp. 38–47.

[6] Borio, D., C. O' Driscoll, and J. Fortuny, "GNSS Jammers: Effects and Countermeasures," *Proc. 6th ESA Workshop on Satellite Navigation Technologies and European Workshop on GNSS Signals and Signal Processing, (NAVITEC)*, Noordwijk, The Netherlands, December 5–7, 2012, pp. 1–7.

[7] Troglia Gamba, M., et al., "FPGA Implementation Issues of a Two-pole Adaptive Notch Filter for GPS/Galileo Receivers," *Proc. 25th Int. Technical Meeting of the Satellite Division of the Institute of Navigation (ION GNSS 2012)*, Nashville, TN, September 17–21, 2012, pp. 3549–3557.

[8] Gao, G. X., "DME/TACAN Interference and Its Mitigation in L5/E5 Bands," *Proc. 20th Int. Technical Meeting of the Satellite Division of the Institute of Navigation (ION GNSS 2007)*, Fort Worth, TX, September 25–28, 2007, pp. 1191–1200.

[9] Hegarty, C., et al., "Suppression of Pulsed Interference Through Blanking," *Proc. 56th Annual Meeting of the Institute of Navigation and of the IAIN World Congress*, San Diego, CA, June 26–28, 2000, pp. 399–408.

[10] Grabowski, J., and C. Hegarty, "Characterization of L5 Receiver Performance Using Digital Pulse Blanking," *Proc. 15th Int. Technical Meeting of the Satellite Division of the Institute of Navigation (ION GPS 2002)*, Portland, OR, September 24–27, 2002, pp. 1630–1635.

[11] De Angelis, M. R., et al., "An Analysis of Air Traffic Control Systems Interference Impact on Galileo Aeronautics Receiver," *Proc. IEEE Int. Radar Conf.*, May 9–12, 2005, pp. 585–595.

[12] Denks, H., A. Steingass, and A. Hornbostel, "GNSS Receiver Testing by Hardware Simulation with Measured Interference Data from Flight Trials," *Proc. 22nd Int. Technical Meeting of the Satellite Division of the Institute of Navigation (ION GNSS 2009)*, Savannah, GA, September 22–25, 2009, pp. 1–10.

[13] Musumeci, L., J. Samson, and F. Dovis, "Experimental Assessment of Distance Measuring Equipment and Tactical Air Navigation Interference on GPS L5 and Galileo E5a Frequency Bands," *Proc. 6th ESA Workshop on Satellite Navigation Technologies and European Workshop on GNSS Signals and Signal Processing (NAVITEC)*, Noordwijk, The Netherlands, December 5–7, 2012, pp. 1–8.

[14] Brown, A., and N. Gerein, "Test Results of a Digital Beamforming GPS Receiver in a Jamming Environment," *Proc. 14th Int. Technical Meeting of the Satellite Division of the Institute of Navigation (ION GPS 2001)*, Salt Lake City, UT, September 11–14, 2001, pp. 894–903.

[15] Konovaltsev, A., et al., "Mitigation of Continuous and Pulsed Radio Interference with GNSS Antenna Arrays," *Proc. 21st Int. Technical Meeting of the Satellite Division of the Institute of Navigation (ION GNSS 2008)*, Savannah, GA, September 16–19, 2008, pp. 2786–2795.

[16] Chen, Y. H., et al., "Real-Time Software Receiver for GPS Controlled Reception Pattern Antenna Array Processing," *Proc. 23rd Int. Technical Meeting of the Satellite Division of the Institute of Navigation (ION GNSS 2010)*, Portland, OR, September 21–24, 2010, pp. 1932–1941.

[17] Kurz, L., et al., "An Architecture for an Embedded Antenna-Array Digital GNSS Receiver Using Subspace-Based Methods for Spatial Filtering," *Proc. 6th ESA Workshop on Satellite Navigation Technologies and European Workshop on GNSS Signals and Signal Processing, (NAVITEC)*, Noordwijk, The Netherlands, December 5–7, 2012, pp. 1–8.

7

Interference Mitigation Based on Transformed Domain Techniques

Luciano Musumeci and Fabio Dovis

7.1 Introduction

Detection and mitigation of RF signals that are interfering with GNSS signals rely on the ability to clearly identify the presence of spurious components and, possibly, to remove them, without damaging the structure of the useful signal. To this purpose, it can be useful to search for a representation of the received signal in a domain different from the classical time and frequency domains introduced in Chapter 6, where useful and spurious contributions can be better isolated. The choice of the best transformation projecting the signal in the new domain depends on the nature of the interfering source, on the chosen mitigation technique, and on the complexity that can be afforded by the signal processing stages of the receiver. Therefore, this chapter will introduce some examples of techniques that have been recently proposed and assessed in the GNSS literature. However, due to the large number of possibilities for invertible transformations studied in the digital signal processing field, other options might be investigated in the future to better cope with different classes of interfering signals. In particular, we introduce three families of transformations, providing a representation of the received digitized signal in the time-frequency, time-scale, and subspace domains. For all of the introduced methods, decomposition stages, detection algorithms, and interference removal processes are presented, together with some case studies demonstrating the potential of each technique.

7.2 Transformed Domain Techniques

Recently, researchers in the GNSS field have started investigating a new family of interference detection and suppression solutions based on the use of advanced signal processing techniques that allow for representation of the signal digitized by the ADC of the receiver, in a different domain, where the information related to the interference can be better identified, isolated, processed, or removed. Such a new family of algorithms will be referred to in the following as transformed domain (TD) techniques, and the different logical steps of the process are summarized in Figure 7.1.

In the analog domain, a signal $x(t)$ can be represented in a transformed domain $X(\alpha, \beta, \gamma, \ldots)$ by exploiting a set of basis functions $h(\alpha, \beta, \gamma, \ldots)$ such that

$$X(\alpha,\beta,\gamma,\ldots) = \langle x(t), h(\alpha,\beta,\gamma,\ldots)\rangle = \int_{-\infty}^{+\infty} x(t)\cdot h^{*}(t,\alpha,\beta,\gamma,\ldots)dt \quad (7.1)$$

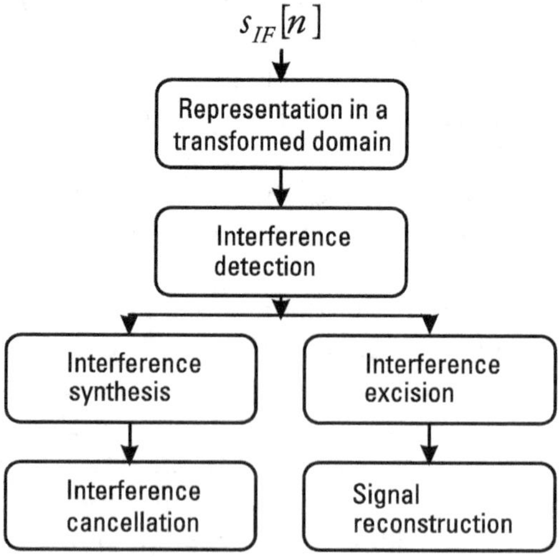

Figure 7.1 Typical TD technique block scheme.

where $X(\alpha, \beta, \gamma, \ldots)$ is the representation of the signal in the transformed domain, and where the set of variables $(\alpha, \beta, \gamma, \ldots)$ represents the dimension of the transformed domain.

In general, the set of functions is discretized by choosing discrete values of $(\alpha, \beta, \gamma, \ldots)$ in order to obtain a set of orthonormal functions $h_k(t, \alpha, \beta, \gamma, \ldots)$. The choice of the basis functions, and thus the decomposition, should allow for identification of the components $X_k(\alpha, \beta, \gamma, \ldots)$ belonging to the interfering signal, thus separating them from the useful components.

The signal is then represented in such a domain by $X_k(\alpha, \beta, \gamma, \ldots)$ weighting the set of basis functions. Thus, the reconstruction of $x(t)$ can be achieved by

$$x(t) = \sum_k X_k(\alpha,\beta,\gamma,\ldots)h_k(t,\alpha,\beta,\gamma,\ldots) \tag{7.2}$$

The majority of the TD techniques that have been investigated in the literature rely on a detection algorithm based on a thresholding operation. According to the Neyman-Pearson criterion [1], the threshold value V_{th} is set considering the null hypothesis H_0 (absence of interference) and a desired false alarm probability, defined as

$$p_{\text{fa}}\left(V_{\text{th}},\alpha,\beta,\gamma,\ldots\right) = P\left(\left|X_k(\alpha,\beta,\gamma,\ldots) > V_{\text{th}}\right|H_0\right) \tag{7.3}$$

where $X_k(\alpha, \beta, \gamma, \ldots)$ is the chosen TD representation.

Basically, the values of $X_k(\alpha, \beta, \gamma, \ldots)$ are compared to a mask that represents the expected GNSS signal representation in the absence of interference. Finally, two options for the interference suppression algorithm can be considered. A synthetic reconstruction of the interfering signal, by means of an antitransformation process based on the identified interference coefficients, can be performed in order to subtract the interferer from the composite received signal (interference cancellation). A different approach can be based on direct suppression in the transformed domain of the interference components believed to belong to the interference, before performing an antitransformation operation for signal reconstruction (interference excision).

It is clear that the chosen transformation must be invertible in order both to be able to generate the synthetic version of the interfering signals in the case of mitigation by cancellation, as well as to reconstruct the "interference-free" GNSS signal in the case of mitigation by excision in the transformed domain.

Due to the typical architecture of modern GNSS receivers, transformations that allow efficient implementation in the digital domain are preferable. In fact, detection/mitigation units based on TD techniques could be implemented in the receiver right after the ADC conversion stage, processing the signal samples before feeding them to the acquisition and tracking stages of the receiver.

7.3 Time-Frequency Representation

A well-known transformation leading to a time-frequency representation of the signal at the ADC output can be achieved by means of the so-called "orthogonal-like Gabor expansion." The transformation proposed in [2] is based on a discretization of the time-frequency plane into a lattice of coordinates $t_n = nT$ and $f_m = m\Omega$ where $-\infty \leq n, m \leq \infty$ and the terms T, Ω represent lattice intervals of, respectively, time and frequency variables. Thus, for an arbitrary signal $x(t)$, the Gabor expansion assumes the form

$$x(t) = \sum_{m}\sum_{n} C_{m,n} h_{m,n}(t) \tag{7.4}$$

where $h_{m,n}(t) = \upsilon(t - mT)e^{jn\Omega t}$ is a Gaussian one-dimensional function allowing the best time-frequency resolution.

The Gabor coefficients $C_{m,n} = C(mT, n\Omega)$ are not unique and they provide information concerning the signal intensity for each point of the discretized time-frequency plane. Reference [2] also shows that the Gabor expansion can be derived from the sampled version of the continuous short-time Fourier transform, since

$$C(mT, n\Omega) = \left\langle x(t), h_{m,n}(t) \right\rangle = \int_{-\infty}^{+\infty} x(t) h_{m,n}^{*}(t)\,dt = \int_{-\infty}^{+\infty} x(t)\upsilon^{*}(t - mT)e^{-jn\Omega t}\,dt \tag{7.5}$$

where $\langle x(t), h_{m,n} \rangle$ is the inner product between the function $x(t)$ and the functions $h_{m,n}(t)$.

Such an orthogonal-like Gabor expansion decomposition has been used in [3] to implement a mitigation technique for chirp signal and linearly frequency-modulated signals. In such a case the detection threshold is set by analyzing the ideal TF representation of the GNSS signal in the absence of interference. Mitigation is achieved by subtraction of the synthetic reconstructed replica of the interfering signal. More details of this TF-based interference mitigation method can be found in [2] and [3].

7.4 Time-Scale Domain: The Wavelet Transform

The wavelet transform (WT) is a well-known technique used in the field of signal processing for different purposes. The first attempts at developing an interference mitigation algorithm based on the use of the WT for GNSS were presented in [4] and [5], addressing the issue of pulsed interference mitigation. In such works the WT is employed to obtain the time-scale representation of the incoming interfered signal.

The WT of a signal provides a representation of the signal components in a domain, spanned by a set of functions that, different from the STFT, can be seen as bandpass filters with a bandwidth decreasing as their central frequency decreases, thus granting a uniform resolution in the decomposition of the signal under analysis.

The basis functions employed in the wavelet transform belong to the set

$$h_k(t) = a^{-k/2} h(a^{-k} t) \tag{7.6}$$

In the frequency domain the functions can be written as $H_k(j\Omega) = a^{k/2} H(ja^k\Omega)$ where $a > 1$ and $k \in Z$.

All of the spectra are obtained by a frequency-scaling operation of a prototype function $H(j\Omega)$. The WT can be implemented by means of a nonuniform filter bank with responses $h_k(t)$. The scale factor $a^{-k/2}$ is introduced as a normalization factor in order to ensure constant energy independent from k, as well as the ratio between the bandwidth and the center frequency Ω_k. Given an arbitrary input $x(t)$, the output of the filter $h_k(t)$ can be computed as

$$w_k(t) = \int_{-\infty}^{\infty} x(t) h_k(\tau - t)\, dt = a^{-k/2} \int_{-\infty}^{\infty} x(t) h\left(a^{-k}(\tau - t)\right) dt \tag{7.7}$$

Furthermore, since the filter bandwidth $H_k(j\Omega)$ is smaller for larger k, its output can be sampled at a lower rate. Equivalently in the time domain, the width of $h_k(t)$ is larger, thus it is possible to move the window by a larger step size [6].

The continuous variable τ can be sampled at na^kT, where n is an integer, to obtain a discrete wavelet transform (DWT). In this way, the step size for the window movement is a^kT, and it increases as the center frequency of the filter Ω_k decreases. Thus, the set of coefficients obtained according to

$$X_{\mathrm{DWT}}(k,n) = a^{-k/2}\int_{-\infty}^{\infty} x(t)h(nT - a^{-k}t)\,dt = \int_{-\infty}^{\infty} x(t)h_k(na^kT - t)\,dt \quad (7.8)$$

is the convolution between $x(t)$ and $h_k(t)$ evaluated at a discrete set of points. Figure 7.2 shows the transfer functions of each branch of the nonuniform filter bank obtained by a dyadic scaling operation ($a = 2$) of the Meyer wavelet function [7]. The family of filters denoted as $h_{kn}(t)$ is the set of the analysis filter.

Perfect reconstruction of the signal $x(t)$ can be achieved by choosing a proper synthesis filter bank. Given a set of wavelet coefficients $X_{\mathrm{DWT}}(k, n)$, the inverse DWT can be achieved according to

$$x(t) = \sum_k \sum_n X_{\mathrm{DWT}}(k,n)\psi_{kn}(t) \quad (7.9)$$

Thus, the signal analysis/synthesis can be achieved by perfect reconstruction of paraunitary quadrature mirror filter (QMF) banks that satisfy the condition $\psi_{kn}(t) = h^*_{kn}(-t)$ [6].

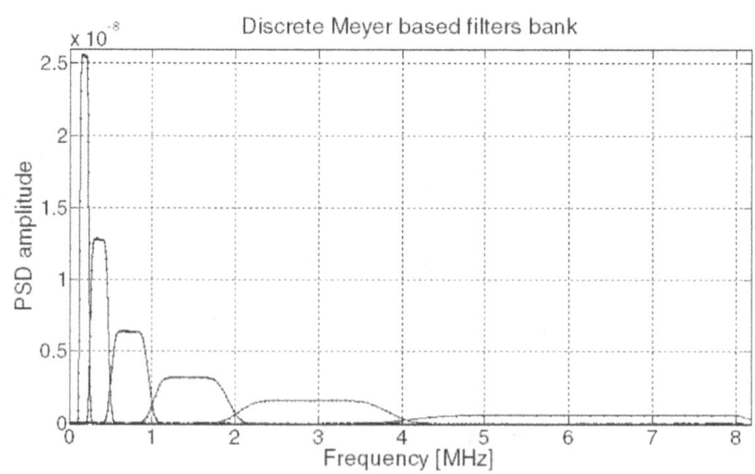

Figure 7.2 Meyer wavelet filter bank responses.

7.4.1 The Discrete Time Wavelet Transform

The relation defined in (7.9) is the DWT since k and n are integer numbers, but it is not the discrete-time since t is continuous. Reference [6] shows that an orthonormal basis function can be generated by a discrete-time QMF bank under certain condition. As an example, let consider the case $a = 2$, known as a dyadic wavelet decomposition, and further assume $T = 1$. Given a paraunitary pair of frequency responses $H(z)$ and $G(z)$, the wavelet function $\psi(t)$ will satisfy the equation

$$\psi(t) = 2^{1/2} \sum_{n \in \mathbb{Z}} h[n]\phi(2t - n) \tag{7.10}$$

where $\phi(t)$ is the so-called "scaling function," which satisfies the scaling equation

$$\phi(t) = 2^{1/2} \sum_{n \in \mathbb{Z}} g[n]\phi(2t - n) \tag{7.11}$$

with $h[n]$ and $g[n]$ being the wavelet vector and the scaling vector, respectively, and derived as the inverse zeta-transform of $H(z)$ and $G(z)$. The scaling function $\phi(t)$ and the wavelet function $\psi(t)$, satisfying (7.12) and (7.13), under the assumption that $H(z)$ and $G(z)$ form a paraunitary filter pair, are shift-orthogonal and span-orthogonal subspaces V_i and W_i, where V_i is spanned by the set $\{2^{i/2}\phi(2^i t - n)\forall n \in Z\}$ while W_i is spanned by the set $\{2^{i/2}\psi(2^i t - n)\forall n \in Z\}$.

From the orthogonality properties of the subspaces spanned by the wavelet and scaling function, the wavelet and scaling vectors must be individually shift-orthogonal and orthogonal to each other. The terms $H(z)$ and $G(z)$ represent the pair QMF that will be employed for the discrete-time DWT. The equivalent expression for (7.6) in the digital domain is $H_k(e^{jw}) = H\left(e^{j2^k\omega}\right) \rightarrow H_k(z) = H\left(z^{2^k}\right)$ where k is a nonnegative integer. Reference [6] shows that $H_k(z)$ is a multiband (rather than a passband) filter. Thus, in order to obtain passband filters, a lowpass filter $G(z)$ is employed. Thus, according to a dyadic scaling operation, the nonuniform filter bank responses at each branch are obtained as

$$H(z), G(z)H(z^2), G(z)G(z^2)H(z^4)\ldots \tag{7.12}$$

The equivalent impulse response of each branch is the digital version of $h_{kn}(t)$.

The orthogonal basis functions $h_{mn}(t)$ in (7.4) employed for the STFT decomposition have equal frequency bandwidths and represent a set of windows in time with equal duration. Such a set of basis functions leads to a different resolution in the characterization of the frequency components of the signal. Many cycles of a high-frequency signal can be captured within the duration $v(t)$, whereas this is not the case for a low-frequency signal. For this reason, the resolution of the STFT is poor at low frequency, but improves as the frequency increases [6]. In fact, the STFT can be seen as a bandpass uniform filter bank in which each filter frequency response has the same bandwidth and a different central frequency.

The filtering operation of the WT can also be iterated at the higher frequency branch of the wavelet decomposition, thus obtaining a uniform filter bank performing the so-called "wavelet packet decomposition" (WPD). The output of each branch provides a set of coefficients (packets) that represent a determined frequency portion of the incoming decomposed signal.

The feature of the wavelet functions through the WT or the WPD that is actually useful when designing the interference detection algorithm is the ability for the functions to be compact in both the time and frequency domains (fast decay). Such a feature allows them to identify local phenomena at different scales, thus identifying the interfering source from the GNSS signals that essentially appears like noise in the time-scale domain.

7.4.2 Wavelet Packet Decomposition Based Mitigation Algorithm

The proposed wavelet-based mitigation algorithm is completely based on the WPD, where the discrete-time signal is passed through a uniform wavelet-based filter bank, as shown in Figure 7.3.

In the WPD, the scaling and shifting process is also iterated at higher frequencies, thus resulting in a uniform filter bank. The algorithm for interference detection and suppression is mainly based on three steps:

1. In the *decomposition phase* the incoming GNSS interfered signal is passed through the uniform filter bank, thus achieving the time-scale representation. The number of wavelet stages to apply for the signal decomposition is a free parameter. Later in this work, the optimal number of wavelet decomposition stages will be assessed with respect to interference spectral characteristics and to GNSS receiver performance at both the acquisition and tracking levels.

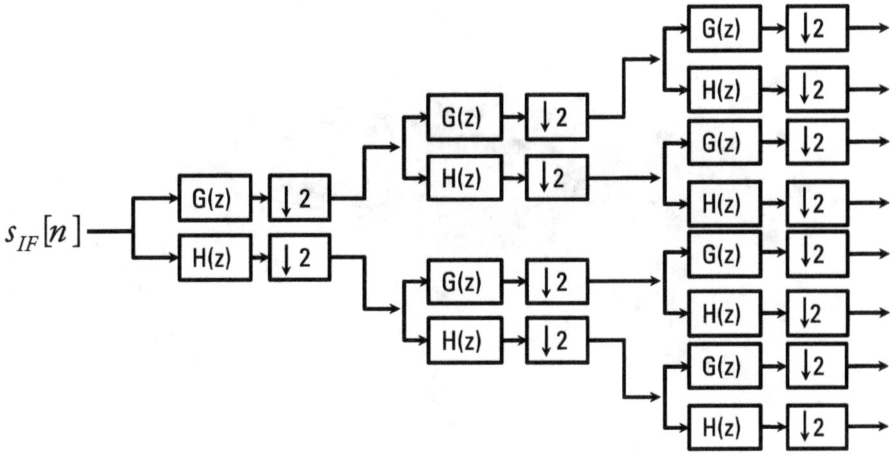

Figure 7.3 Wavelet packet decomposition.

2. The *detection-mitigation phase* is performed in each scale obtained at the output of the filter bank. Interference excision is performed by applying a blanking operation on the time-series of the coefficients. Such a process is based on the suppression of the coefficients in each scale crossing a determined blanking threshold level.
3. The *reconstruction phase* is achieved through an inverse wavelet transform starting from those scales modified after suppression of the interference coefficients.

As an example, Figure 7.4(a) shows the time-scale representation of 10 ms of simulated GPS C/A code signal subject to narrowband interference (NBI). The interfered GPS C/A code signal is simulated by means of a fully software GNSS signal generator. The time-scale floor, after four stages of WPD, shows how the NBI components, located in the determined scales, clearly emerge from the time-scale noise floor, thus easing the detection process.

The black floor in Figure 7.4(a) represents the unique blanking threshold employed for interference coefficient excision. The excision process, based on the suppression of all coefficients in each wavelet packet exceeding the predetermined blanking threshold, leads to a new set of wavelet packets, as shown in Figure 7.4(b). Then signal reconstruction is achieved through a uniform filter bank matched with the uniform filter bank employed for the signal decomposition (Figure 7.3), as mentioned in [6]. Figure 7.5 shows a comparison between the PSD of the simulated interfered GPS C/A code signal before

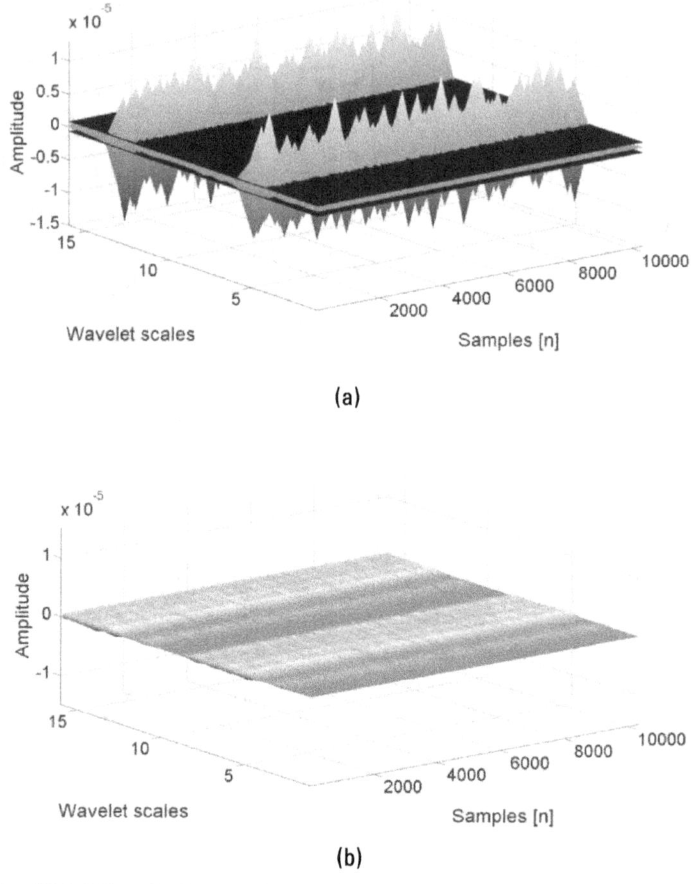

Figure 7.4 GPS C/A code time-scale domain in the presence of narrowband interference (a) before the excision and (b) after the excision.

(black line) and after (gray line) the WPD-based processing. It can be clearly seen that the NBI components are almost completely removed thanks to the selective effect provided by the WPD process, which helps in extracting and separating the interference components from the useful GNSS signal.

7.4.3 WPD-Based Method: Parameter Tuning

The WPD-based method can be tailored to different scenarios of interference. In fact, several parameters can be user defined, and they will be investigated in this section with respect to the mitigation of a NBI.

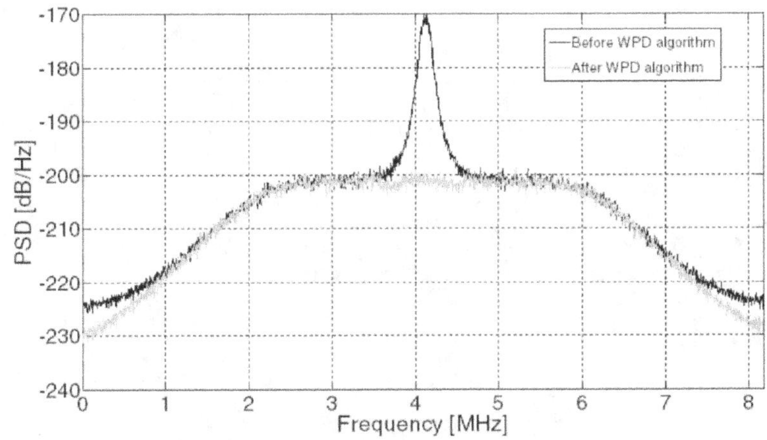

Figure 7.5 Time-scale domain after interference coefficient excision.

Several NBI scenarios have been considered, and a parametric study with respect to the interference bandwidth B_{int}, interference carrier frequency f_{int}, and number of wavelet decomposition stages N has been performed. A fully software GNSS receiver is employed and the best parameters configuration, such as the number of decomposition stages N, is shown.

Wavelet Decomposition Depth N

This first analysis was devoted to the study of the impact of the number of wavelet decomposition stages with respect to the NBI suppression performance. Three different interference scenarios have been considered, combining GPS L1 C/A code signals with NBI 200 kHz far from the intermediate frequency, and results are shown in Figure 7.6.

Here the trend of the acquisition metric $\alpha_{mean} = R_p/M_c$, defined as the ratio between the main acquisition peak and the correlation noise floor, versus the number of wavelet decomposition stages is shown. Acquisition performance is achieved using 1 ms of coherent integration time, and 20 noncoherent accumulations. The three lines are referred to three different interference scenarios characterized by the presence of NBI with respectively 40, 80, and 120 kHz of bandwidth. Increasing the number of WPD stages increases the wavelet scale resolution and thus its frequency selectivity. In all three interference scenarios, increasing N provides better performance in capturing and isolating the NBI components, which in turn means better interference suppression without removing useful signal components, as shown from the increasing trend of α_{mean}. However, a saturation effect can be observed for higher values of N

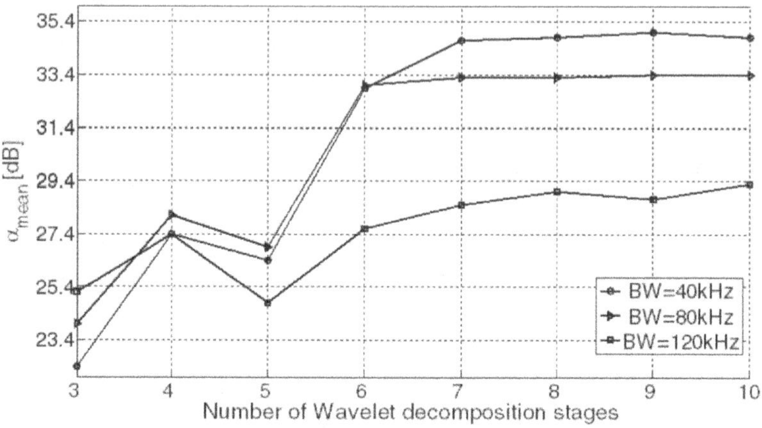

Figure 7.6 Acquisition metric versus WPD depth.

(greater than 7). In such a region, acquisition performance does not improve further since the wavelet packet resolution is already comparable or narrower than B_{int}. Moreover, as expected, performance of such a technique is limited and for larger B_{int}, lower acquisition metric values are achieved.

Wavelet Families Comparison

In Section 7.3 a Meyer wavelet was introduced. Several other wavelet functions exist, and they have different spectral features when used to generate the filters responses for the WPD (see, for example, [8]). Further analyses have focused on the use of different wavelet functions. In particular, a wavelet function derived from the orthogonalization process of a Gaussian function (modified Gaussian wavelet; [9]) shows good performance for adopted interference mitigation. Figure 7.7 depicts the wavelet filter bank obtained from the Gaussian wavelet function. Note that such filters are characterized by a more frequency selective response.

Figure 7.8 shows the acquisition metrics α_{mean} with respect to the number of wavelet decomposition stages when mitigating NBI on a carrier frequency 200 kHz far from the intermediate frequency and with a bandwidth of 120 kHz. The square marked line refers to the acquisition metric achieved when the NBI mitigation is performed by means of the time-scale decomposition based on modified Gaussian derived wavelet functions; the circle marked line refers to the acquisition metric achieved by exploiting Meyer-derived wavelet functions.

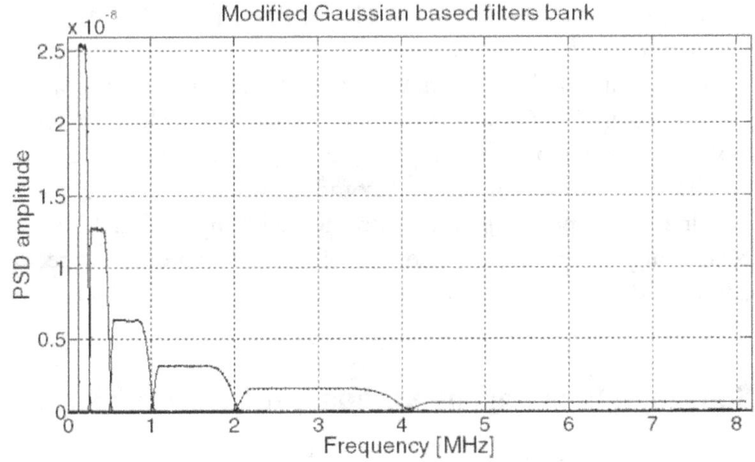

Figure 7.7 Modified Gaussian wavelet filter bank response.

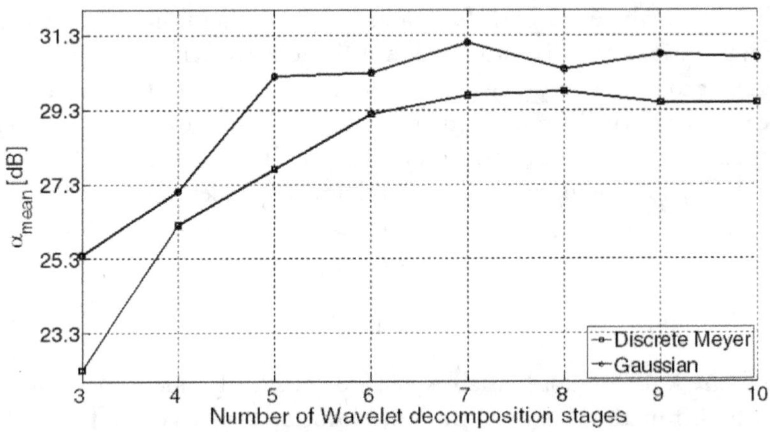

Figure 7.8 Acquisition metric: Gaussian versus Meyer wavelet.

7.4.4 Computational Complexity

Although a wavelet-based mitigation algorithm is good at suppressing interference components, its implementation is characterized by a complexity that is not negligible. The computational burden is mainly determined by the number of wavelet decomposition stages N, which determines the number of filtering operations according to the exponential law 2^N. Furthermore, the same number of filtering operations is employed for signal reconstruction purposes.

All filtering operations are realized with FIR filters of length L. Each output sample is obtained with L products and one single sum; thus, the total number of performed operations for decomposition and reconstruction of n samples of an incoming signal is $O(n, N, L) = 2 \cdot 2^N \times (nL + n)$. However, the filter bank implementation allows for sample-by-sample processing of the incoming signal, at the price of a delay at the decomposition stage and by the reconstruction filter bank operating on the thresholded samples. Furthermore, the wavelet-based algorithms can represent an efficient postprocessing technique for interference detection and characterization.

7.5 Subspace Domain: The Karhunen-Loève Transform

The Karhunen-Loève transform (KLT) provides a decomposition of the signal in a vectorial space using orthonormal functions that in principle can have any shape, different from the other transforms, as for instance in the Fourier transform where the basis functions are sinusoidal functions. Its use for space applications was first proposed in [10]. Here, the KLT is employed as an instrument to detect very weak signals hidden in noise, in the framework of the Search for Extra Terrestrial Intelligence program. However, in [11] a first attempt at CWI detection based on the use of KLT was presented.

The KLT decomposition of a general time-dependent function is given by

$$x(t) = \sum_{n=1}^{\infty} Z_n \phi_n(t) \tag{7.13}$$

where Z_n are scalar random variables that are statistically independent and $\phi_n(t)$ are the basis functions, derived from the covariance matrix of a digitized version of the stochastic process of $x(t)$. The KLT offers better separation between the deterministic components within the received signal and the stochastic ones. Unlike the basis functions, which represent the behavior in time of the signal to be decomposed, the random variables Z_n are obtained projecting the given stochastic process $x(t)$ over the corresponding eigenvector $\phi_n(t)$; that is,

$$Z_n = \sum_{-\infty}^{+\infty} x(t) \phi_n(t) dt \tag{7.14}$$

References [11] states that the Karhunen-Loève expansion is the only possible statistical expansion in which all of the expansion terms are uncorrelated from each other. The nature of the KLT is independent of the specific kind of interfering signal, thus making the KLT capable of successfully detecting not only CWI, but also NBI, WBI, and chirp interference.

7.5.1 KLT Interference Detection and Suppression Algorithm

The KLT decomposition has been implemented according to the following steps:

- Computation of the Toeplitz matrix of the autocorrelation of the received signal;
- Determination of the eigenvalues of the Toeplitz matrix and of the related eigenfunctions
- Determination of the Z_n coefficients determination according to (7.14).

Figure 7.9 shows the capability of the KLT in separating the deterministic and stochastic components within the signal to decompose. The KLT decomposition is achieved solving the eigenvalues[1] problem for the Toeplitz matrix of the autocorrelation function of 100 μs of simulated GPS C/A code signal at nominal power in two cases: (1) in an interference-free environment and (2) when interfered with by an NBI signal (10 kHz) centered on the intermediate frequency with a power equal to −120 dBW.

Figure 7.9 reports the trend of the normalized eigenvalues λ and the Z_n coefficients obtained from the KLT decomposition. Note that the distribution of the eigenvalues suggests a method for detecting interference. In fact, when the interference is present, a small number of eigenvalues which have a greater magnitude with respect to the others (bottom plot), which is different from the case for an interference-free environment (top plot).

A detection method based on the eigenvalues magnitude observation is proposed in [12]. Basically, the highest magnitude eigenvalues, which represent the interference components, are detected and an inverse KLT is applied considering only the eigenfunctions representative of the noise in which the GNSS component is embedded. To define a threshold value based on analytical justifications, the statistical distribution of the Z_n coefficients for a signal in an interference-free environment should be studied. In [10] the authors state that the distribution of the KLT coefficients for a unitary stationary white noise is Gaussian. However, this is not the case, because even if the GNSS signal

[1] A scalar λ is called an eigenvalue of the $n \times n$ matrix $\mathbf{A}$ if there is a nontrivial solution $\mathbf{x}$ of $\mathbf{Ax} = \lambda\mathbf{x}$. Such an $\mathbf{x}$ is called an eigenvector corresponding to the eigenvalue λ.

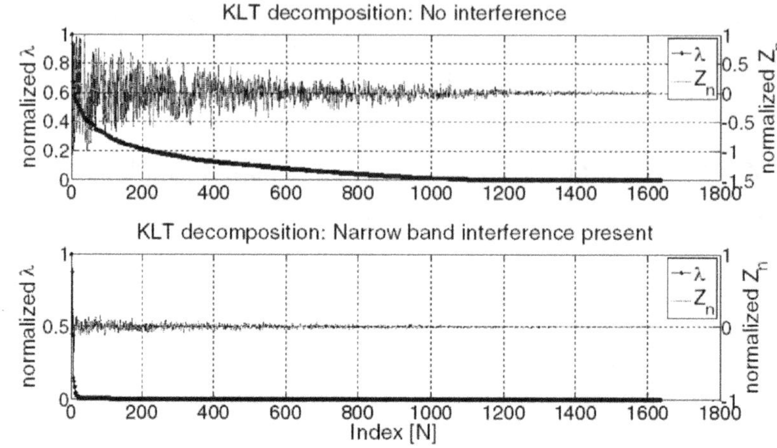

Figure 7.9 KLT decomposition comparison between an interference-free case (top) and a case in which interference is present (bottom).

is completely buried in the noise, some deterministic components due to the GNSS codes are contained in it. Furthermore, it would be desirable to have a method that is independent of the interference features. Thus, an energy-based detection algorithm is developed, analyzed, and proposed in [13], and it will briefly described in Section 7.6.2 where the application of the KLT-based method for pulsed interference removal is presented.

The KLT-based method offers good performance in extracting interference information from a received signal, but the computational burden of its implementation is quite heavy since an eigenvalue problem has to be solved.

7.6 Case Study: A Pulsed Interference Environment

To assess the interference suppression capabilities of both the WPD- and KLT-based methods and their advantages with respect to the use of a traditional pulse blanking operation, an example of their applications for pulsed interference removal is presented here.

A particular pulsed interference scenario, typical of an aviation environment, is considered. Figure 7.10 shows the 10-ms spectrum of simulated Galileo E5a signals interfered by a set of simulated pulsed interfering signals, typically originated by DME beacons located near the airports. Such pulsed interfering signals appear in the spectrum as narrowband

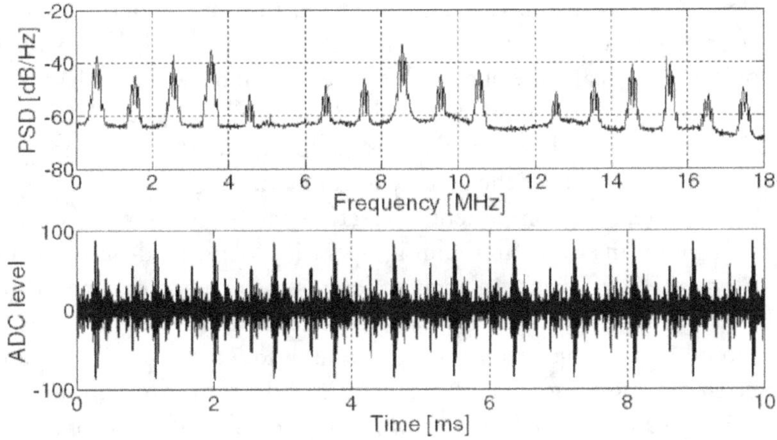

Figure 7.10 Galileo E5a power spectral density in the presence of pulsed interference.

interference jamming the entire GPS L5 and Galileo E5a frequency bands, as mentioned in Chapter 2.

7.6.1 WPD Applied to Pulsed Interference

An example of time-scale representation of the signal at the ADC output is shown in Figure 7.11, where five stages of WPD have been employed on the IF samples of the pulsed interfered data set shown in Figure 7.10.

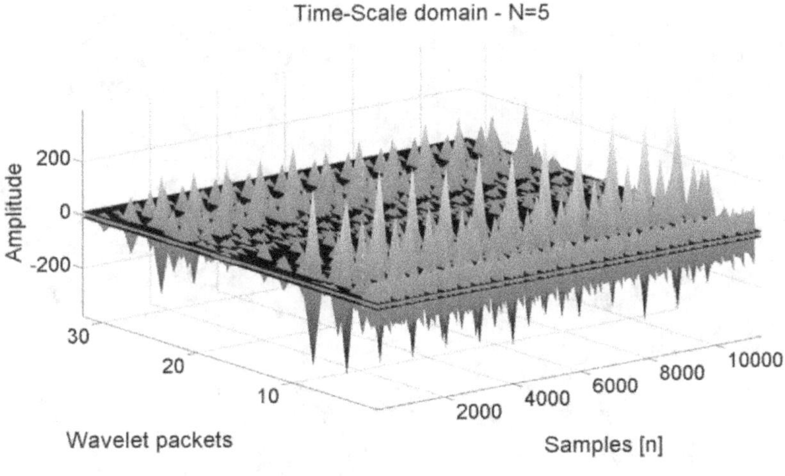

Figure 7.11 Time-scale representation.

Meyer [7] wavelets have been considered in order to derive all of the filter frequency responses employed in the uniform filter bank. After five stages of WPD, 32 scales are obtained, each of which represents a determined frequencies region of the interfered received Galileo E5a signal. As shown in Figure 7.11, the total pulsed interfering signal reaching the user antenna has components spread all over the time-scale domain.

The interference components detection in the time-scale domain is performed by means of a thresholding operation performed on each single scale. All of the coefficients in each scale crossing a predetermined threshold are then suppressed. Such a process of detection-excision, which is very similar to a blanking operation in each scale, leads to direct suppression of the interference components in the transformed domain. An example of this operation is illustrated in Figure 7.12 where the set of coefficients obtained at the output of a generic branch of the WPD filter bank is shown before and after the blanking.

The same detection criteria based on the Neyman-Pearson criteria are adopted for the determination of the blanking threshold V_{th} to apply in each scale. However, in this example, a unique value of V_{th} is employed for the entire time-scale domain. In fact, since the GNSS signal is completely buried in the noise at the user antenna level, under a Nyquist sampling condition, the filtered digitized noise can be considered to still be uncorrelated, thus allowing the assumption that at the ADC output, the samples in an interference-free environment are still Gaussian distributed with zero mean and variance σ^2. The digitized signal is then processed by the WPD filter bank, made by the

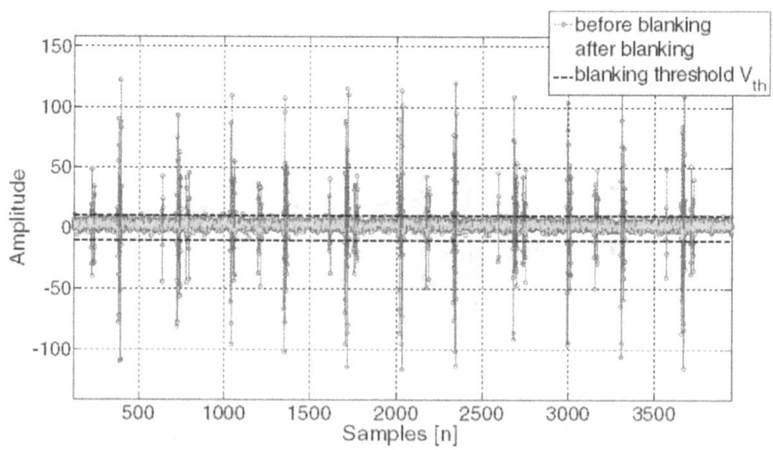

Figure 7.12 Pulse blanking applied on a single scale.

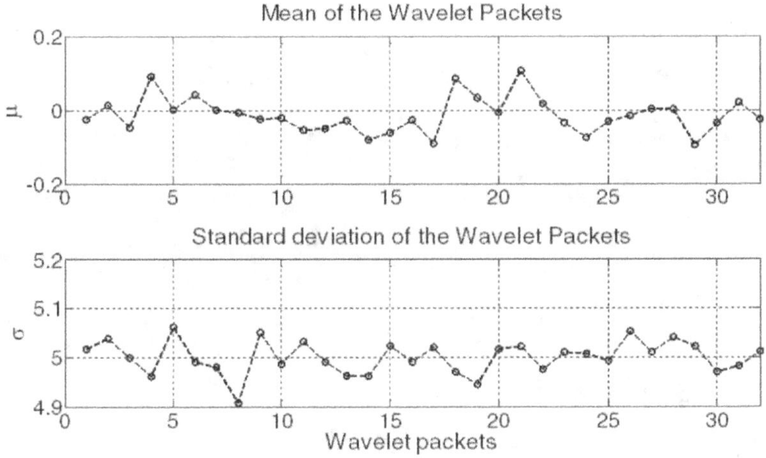

Figure 7.13 Wavelet packet statistical analysis.

filters' responses, which are orthonormal to each other. Thus, the samples at the output of each branch of the filter banks in the absence of interference can be assumed to still be Gaussian distributed with zero mean and variance σ^2. As an example Figure 7.13 shows the mean and the standard deviation values of the wavelet packets achieve after the five stages of WPD applied to the Galileo E5a-Q signal in the absence of interference and in the presence of a flat front-end.

The statistics for all 32 scales are quite similar with a variation among the scales of less than 10%. Only of the scales representing the frequency regions outside the filter cutoff frequencies deviate from this value, but they can be neglected since, if well designed, the front-end bandwidth has a flat frequency response at the GNSS frequency bands. Thus, it is possible to assume a single blanking threshold to be applied to the overall time-scale plane, avoiding the need to define different thresholds for each scale.

The coefficients in the time-scale domain exceeding the blanking threshold are suppressed, as shown in Figure 7.14. Such modified scales are fed to a wavelet based antitransformation block that is in charge of the signal reconstruction.

The benefits of this algorithm can be observed by looking at the spectrum achieved after mitigation (see Figure 7.15). The several pulsed interfering signals appearing as narrowband interference are highly suppressed. Furthermore, unlike a common interference mitigation technique performed in the time domain, as the pulse blanking, where useful signal components

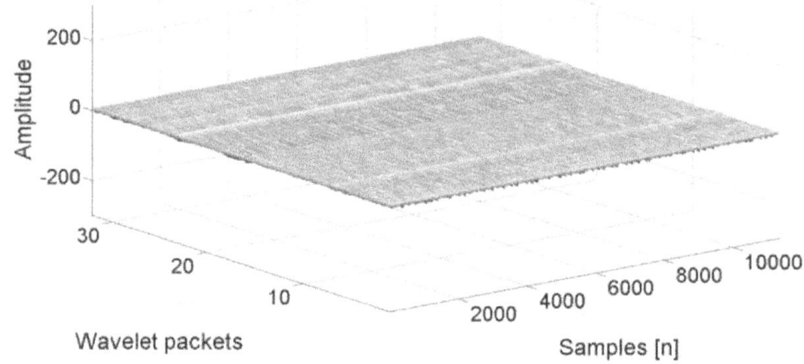

Figure 7.14 Time-scale after interference excision.

are suppressed together with interference, the majority of the useful GNSS signal power is saved, as confirmed by the absence of drops in the spectrum. The main advantages of this algorithm with respect to the Gabor expansion-based algorithm is that no signal storage for the signal decomposition and no synchronization operation during signal reconstruction are needed.

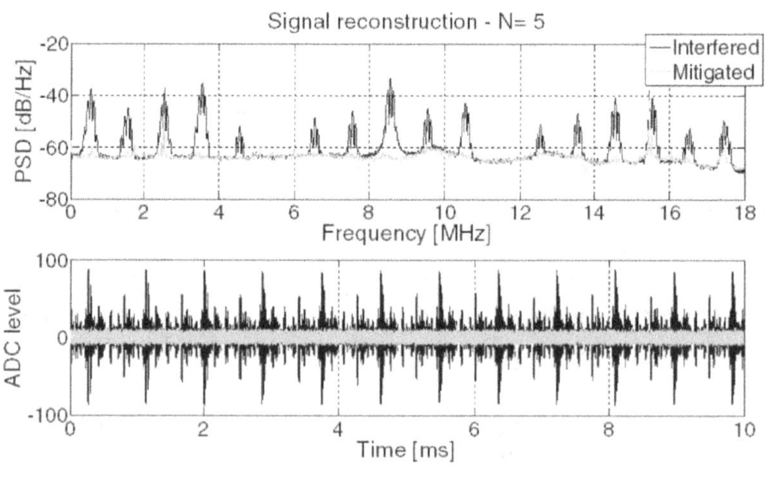

Figure 7.15 Power spectral density before and after the WPD-based interference mitigation.

7.6.2 KLT Applied to Pulsed Interference

KLT-based decomposition and signal reconstruction have been implemented on the pulsed interfered data shown in Figure 7.10. Since the software implementation of these steps requires great computational capabilities, KLT decomposition is performed on small slices, the duration of which is about 16 μs. Figure 7.16 shows the KLT coefficient trend and the total energy of the reconstructed signal when the highest magnitude KLT coefficients up to N are not considered for the reconstruction.

As mentioned earlier, the criterion used for the determination of the number of Z_n coefficients to exclude is based on the signal energy analysis. The intersection point between the dashed curve, which represents the energy of the reconstructed signal when excluding the first N highest Z_n coefficients, and the ideal GNSS signal energy threshold in an interference-free environment (nominal energy) provides the optimal number of highest Z_n coefficients to be suppressed. Following this criterion, the number of KLT coefficients excluded is such that the reconstructed signal energy is about the ideal energy of the signal in an interference-free environment. Operating according to this criterion, the first 20 highest eigenvalues are excluded from the signal reconstruction.

Figure 7.17 shows a comparison of the PSD of the received signal before and after KLT-based pulsed interference excision. From this result we can see that the KLT-based method, like the WPD based algorithm, also offers high-performance abilities in detecting, isolating, and suppressing the dominant deterministic components, which are usually related to the interfering signal,

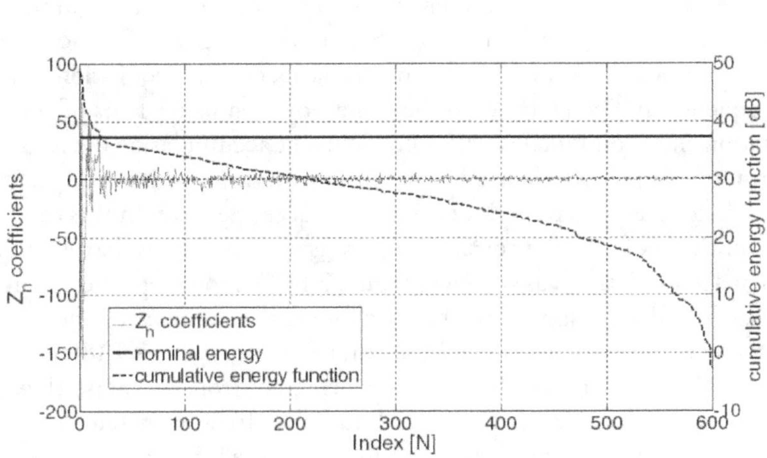

Figure 7.16 KLT decomposition and signal energy.

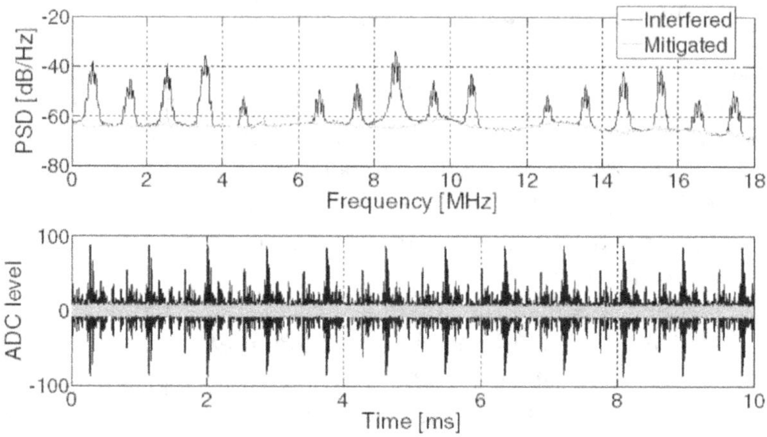

Figure 7.17 Power spectral density comparison before and after KLT-based interference suppression.

contained in the received signal, without causing large distortion of the useful GNSS signal.

7.6.3 TD Techniques Versus Pulse Blanking: Performance Comparison

Figure 7.18 shows the acquisition search spaces of the Galileo E5a pilot channel (PRN 20) obtained in different scenarios. In particular, Figure 7.18(a) shows the acquisition performance of the software receiver when no interference countermeasure is adopted. In this scenario, correct acquisition of Doppler frequency and code delay can be achieved when using 1 ms of coherent integration time combined with 80 noncoherent accumulations. Acquisition performance improves when a simple pulse blanking operation in the time domain is adopted as a pulsed interference countermeasure. In this case, correct acquisition of the true correlation peak is already achieved after 10 noncoherent accumulations, as shown in Figure 7.18(b). However, when using the WPD- or KLT-based algorithm for pulsed interference suppression, acquisition performance improves considerably, as can be seen in Figures 7.18(c) and (d). In both cases, the correlation peak clearly emerges from the noise floor and crosses the acquisition threshold (black floor) after 10 noncoherent accumulations, which are not enough for the acquisition of the interfered data set when no interference countermeasure is applied.

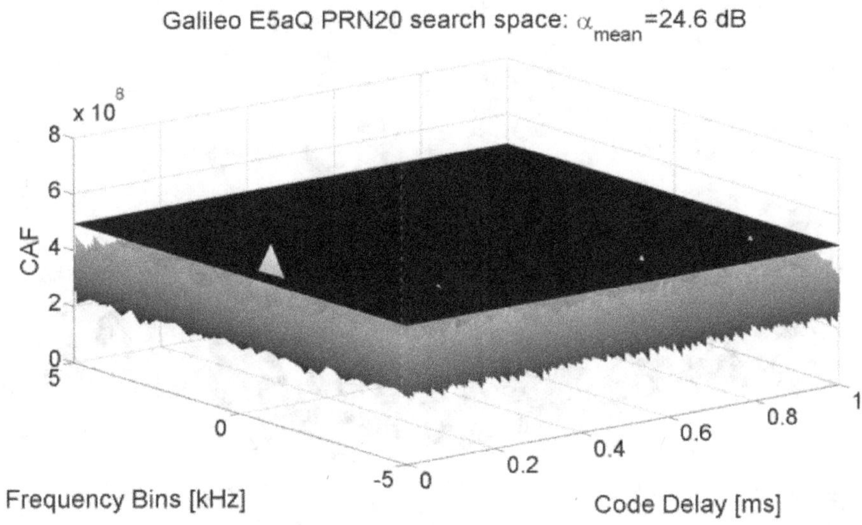

(a)

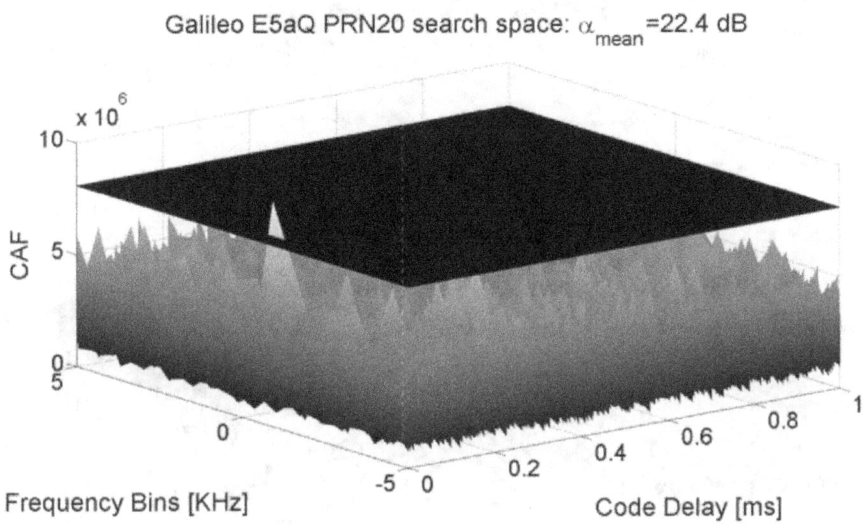

(b)

Figure 7.18 Acquisition search space: (a) no countermeasures; (b) after pulse blanking.

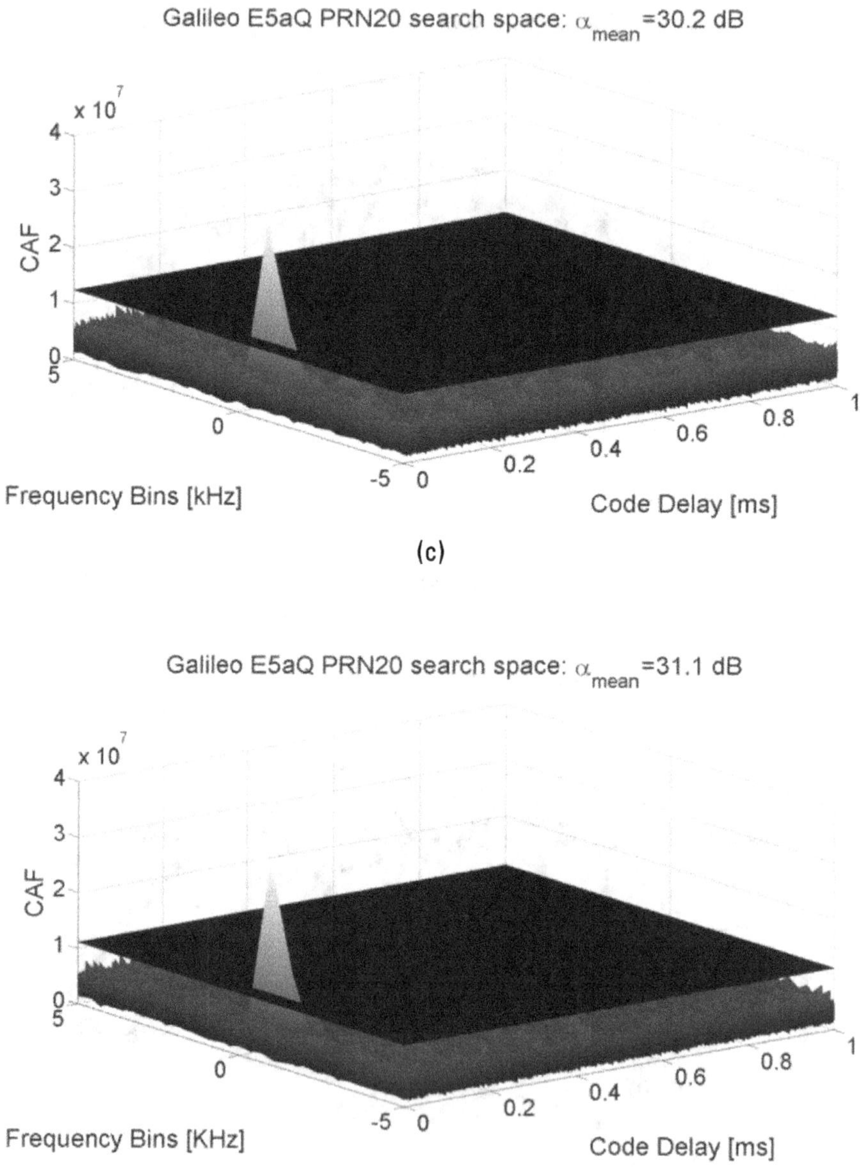

(d)

Figure 7.18 *(Continued)* (c) after WPD-based mitigation; (d) after KLT-based mitigation.

Table 7.1
Acquisition Performance Comparison

Scenario	Noncoherent Accumulations K	α_{mean} (dB)
Interference-free	10	32
Pulsed interfered	80	24.6
After pulse blanking mitigation	10	22.4
After WPD-based mitigation	10	30.2
After KLT-based mitigation	10	31.1

A summary of the acquisition performance is reported in Table 7.1, which shows the acquisition metrics α_{mean} in all four cases considered. We can see clearly how such advanced signal processing algorithms provides higher pulsed interference suppression, resulting in acquisition spaces where the separation between the correlation peak and the noise floor is higher with respect to the case related to the use of a simple blanking operation. Concerning the tracking stage, estimated C/N_0, early-prompt-late correlations, and noise on the data demodulation are analyzed for 10s of Galileo E5a data channel (PRN 20) tracking and reported, respectively, in Figures 7.19, 7.20, and 7.21 for the four scenarios considered for the acquisition performance analysis.

Concerning the estimated C/N_0, note that such advanced signal processing techniques provide almost complete interference component suppression with negligible distortion of the useful GNSS signal components. In

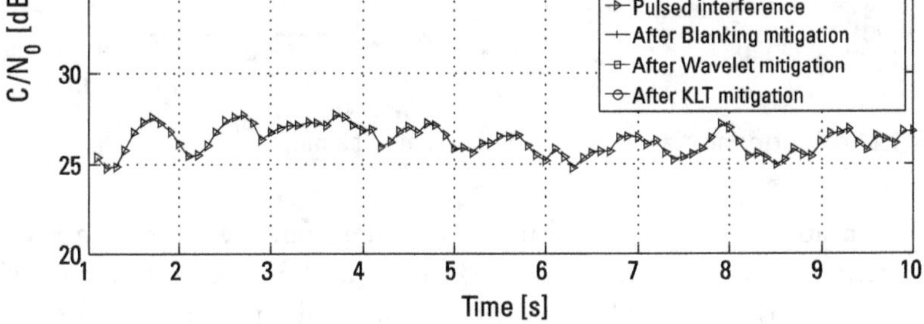

Figure 7.19 Carrier–to–noise density ratio comparison.

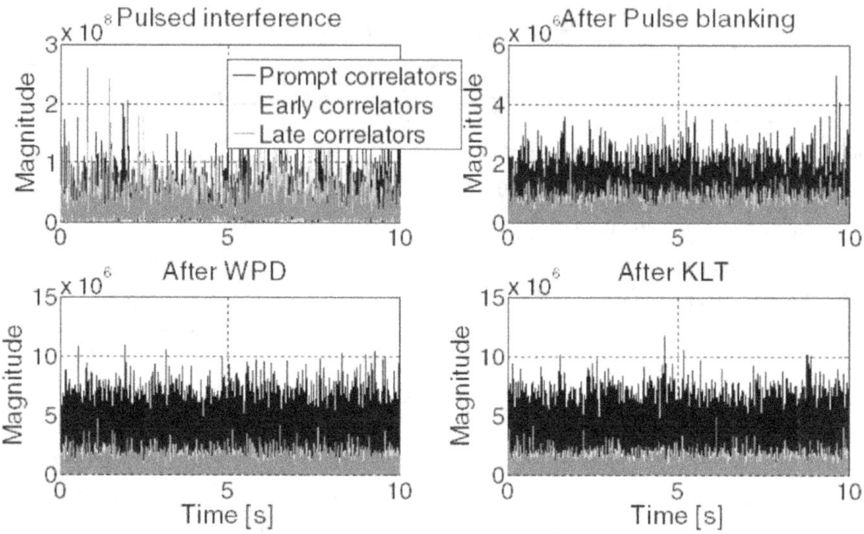

Figure 7.20 Early-prompt-late correlators outputs.

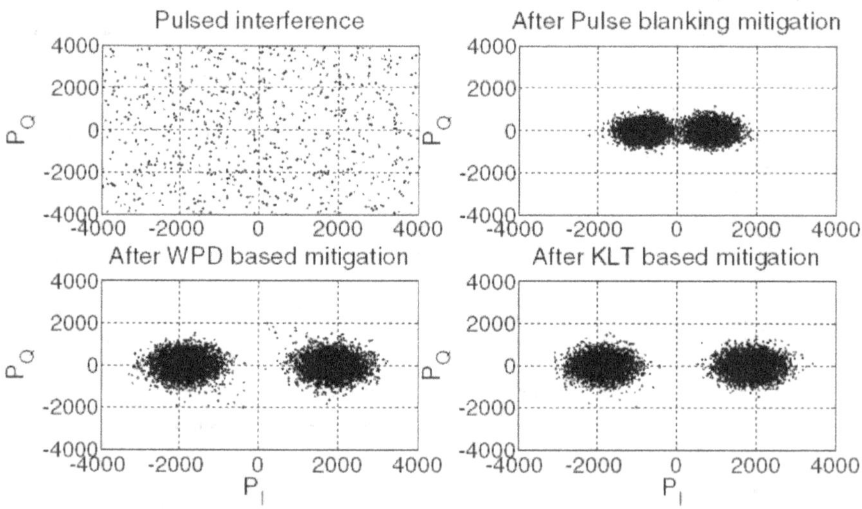

Figure 7.21 Scattering diagram of the demodulated signal.

fact, when adopting pulse blanking as an interference countermeasure, the C/N_0 estimated by the software receiver is around 36.1 dB-Hz while, when adopting both transformed domain techniques, about 4 dB of gain can be observed. The same conclusion can be drawn by looking at Figure 7.20 where

Table 7.2
Acquisition Performance

Scenario	C/N_0 (dB-Hz)	σ_{DLL} (cm)
Pulsed interfered	26.6	—
After pulse blanking mitigation	36.1	76.5
After WPD-based mitigation	40.1	72.6
After KLT-based mitigation	40.6	73

the outputs of the early-prompt-late correlators are depicted in the absence of interference countermeasures, after the application of a simple pulse blanking operation, after WPD-based mitigation, and after KLT-based mitigation. In particular, the prompt correlation amplitude distance from the early and late correlations is higher for the WPD and KLT implementations than for the pulse blanking case, as can be clearly observed in Figure 7.20. Furthermore, in Figure 7.21 the I-Q scattering diagram is noisier in the case of pulse blanking implementation than for the WPD- and KLT-based methods. These results were obtained by setting a predetection integration time T equal to 1 ms and choosing a loop bandwidth equal to 2 and 15 Hz, respectively, for the DLL and PLL. A summary of the software receiver tracking performance is provided in Table 7.2, where the average estimated C/N_0 and DLL jitter during tracking operations are shown.

7.7 Transformed Domain Techniques: Possible Implementation

A discussion concerning the computational complexity required by the implementation of these advanced signal processing techniques for interference removal was provided in Section 7.4.4. The WPD- and KLT-based methods overperform the current interference mitigation algorithm in terms of interference detection and useful GNSS signal distortion. However, their implementation requires more resources with respect to the implementation of standard interference mitigation solutions such as pulse blanking or notch filtering. In particular, the idea of exploiting a KLT-based engine for real-time interference mitigation still seems not to be realistic due to the computational burden required to solve the eigenvalue problem. On the other hand, wavelet transformation is already being implemented in some multimedia data compression techniques, thus its realization for real-time interference mitigation

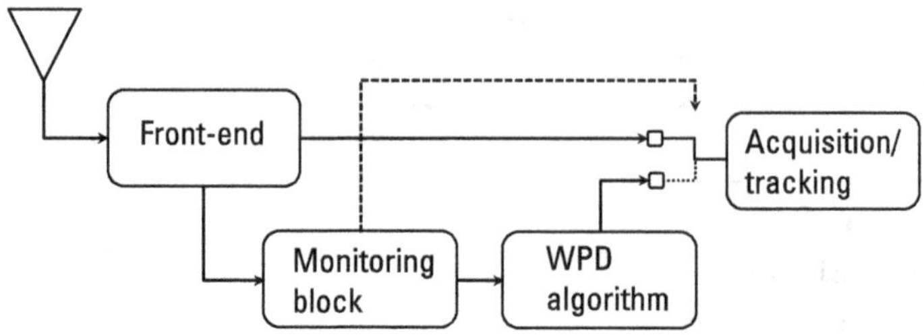

Figure 7.22 Block scheme of WPD-based algorithm for interference removal in a GNSS receiver.

purposes in GNSS receivers can be considered feasible by exploiting powerful FPGA and enhanced microprocessors.

The block scheme shown in Figure 7.22 shows a possible implementation of the WPD algorithm for interference removal in a GNSS receiver. In addition to the engine providing decomposition, detection, and reconstruction, denoted as the WPD algorithm, a signal monitoring block is needed. Such a monitoring block is in charge of analyzing the quality of the received GNSS signal at the ADC output in order to reveal the presence of potential impairments. Several monitoring techniques based on statistical analysis of the received signal pre- or postcorrelation are discussed in the literature and summarized in Chapter 5. Once the monitoring block reveals the presence of an interfering signal, it commutes the input of the acquisition/tracking from the output of the front-end to the output of the WPD algorithm block. Of course, the WPD processing introduces a delay due to the decomposition and reconstruction filter banks. However, the thresholding operates sample by sample, avoiding the use of memory to store data sets of signal samples, as is required by other transformation methods.

7.8 Conclusions

This chapter has introduced some examples of interference mitigation techniques based on the representation of the signal in new domains. The time-frequency transformation provides a good and intuitive representation of time-varying interferers, such as the frequency-modulated jammers. The WPD- and KLT-based algorithms prove to be good performance algorithms for interference detection and suppression. In both cases, representation of the incoming interfered received GNSS signal in a different domain allows for the removal of the interference by excision of portions of the transformed domain, causing negligible distortion of the GNSS useful signal, as demonstrated by looking at the receiver performance achieved at both the acquisition and tracking stages. However, the total computational burden required for their implementation is obviously higher than the complexity foreseen, for example, by a simple pulse blanking implementation or notch filtering design. As already mentioned, concerning the WPD, its complexity is mainly determined by the number of wavelet decomposition stages N, which determines the number of filtering operations according to the exponential law 2^N. At any rate, the presence of smarter algorithms in the literature that provide WPD with a complexity that is logarithmic with the number of wavelet decomposition stages N may represent a solution for the complexity required for its real-time implementation in a GNSS receiver. In contrast, real-time KLT implementation is still a challenging operation due to the fact that such a decomposition foresees the solution of an eigenvalues problem.

References

[1] Kay, S. M., *Fundamentals of Statistical Signal Processing: Detection Theory*, Upper Saddle River, NJ: Prentice Hall, 2010.

[2] Gabor., D., "Theory of Communication," *J. of the IEE*, Vol. 93, No. 3, 1946, pp. 429–457.

[3] Savasta, S., L. Lo Presti, and M. Rao, "Interference Mitigation in GNSS Receivers by a Time-Frequency Approach," *IEEE Trans. on Aerospace and Electronic Systems*, Vol. 49, No. 1, 2013, pp. 415–438.

[4] Anyaegbu, E., et al., "An Integrated Pulsed Interference Mitigation for GNSS Receivers," *Journal of Navigation*, Vol. 61, 2008, pp. 239–255.

[5] Paonni, M., et al., "Innovative Interference Mitigation Approaches, Analytical Analysis, Implementation and Validation," *Proc. 5th ESA Workshop on Satellite Navigation Technologies and European Workshop on GNSS Signals and Signal Processing (NAVITEC)*, Noordwijk, The Netherlands, December 8–10, 2010, pp. 1–8.

[6] Vaydianathan, P. P., *Multirate Systems and Filter Banks*, Upper Saddle River, NJ: Prentice Hall, 1993.

[7] Meyer, Y., *Wavelets Algorithms and Applications*, Philadelphia, PA: Society for Industrial and Applied Mathematics, 1993.

[8] Dovis, F., "Wavelet Based Designed of Digital Multichannel Communications Systems," Ph.D. Thesis, Politecnico di Torino, Italy, 1999.

[9] Dovis, F., M. Mondin, and F. Daneshgaran, "The Modified Gaussian: A Novel Wavelet with Low Sidelobes with Applications to Digital Communications," *IEEE Communications Letter*, Vol. 2, No. 8, 1998, pp. 208–210.

[10] Maccone, C., "The KLT (Karhunen-Loève Transform) to Extend SETI Searches to Broad-band and Extremely Feeble Signals," *Acta Astronautica*, Vol. 67, No. 11–12, 2010, pp. 1427–1439.

[11] Szumski, A., "Finding the Interference: Karhunen-Loève Transform as an Instrument to Detect Weak RF Signals," *Inside GNSS*, No. 3, 2010, pp. 56–64.

[12] Musumeci, L., and F. Dovis, "A Comparison of Transformed-Domain Techniques for Pulsed Interference Removal on GNSS Signals," *Proc. Int. Conf. on Localization and GNSS (ICL-GNSS)*, Sternberg, Germany, June 25–27, 2012, pp. 1–6.

[13] Dovis, F., L. Musumeci, and J. Samson, "Performance Comparison of Transformed-Domain Techniques for Pulsed Interference Mitigation," *Proc. 25th Int. Technical Meeting of the Satellite Division of the Institute of Navigation (ION GNSS 2012)*, Nashville, TN, September 17–21, 2012, pp. 3530–3541.

[14] Musumeci, L., J. Samson, and F. Dovis, "Experimental Assessment of Distance Measuring Equipment and Tactical Air Navigation Interference on GPS L5 and Galileo E5a Frequency Bands," *Proc. 6th ESA Workshop on Satellite Navigation Technologies and European Workshop on GNSS Signals and Signal Processing (NAVITEC)*, Noordwijk, The Netherlands, December 5–7, 2012, pp. 1–8.

8

Antispoofing Techniques for GNSS

Marco Pini and Davide Margaria

8.1 Introduction

GNSSs enable a variety of location-based services (LBSs), and the use of satellite navigation receivers is increasing in several fields of application. The mass-market sector drives the use of navigation technologies thanks to the wide spread of smartphones and tablets with embedded GNSS chipsets. However, there is also growing demand for high-quality LBSs that are able to provide highly accurate and reliable measurements in other fields of application. As shown in previous chapters, interfering signals, either intentional or from other wireless systems, cause poor navigation performance, low positioning accuracy and, in severe conditions, a complete loss of signal tracking. Currently, there is a tremendous need to design GNSS receivers that are robust to interference because they are often employed in more complex systems. For example, in the maritime sector when a ship loses the GNSS signals, multiple systems fail simultaneously, namely, the automatic information system (AIS) transponder, the ship's gyro calibration system, and the digital selective calling system [1].

Indeed, a new consciousness has arisen since it has become evident that surveillance and safety-critical systems that rely on GNSS (e.g., dangerous goods transportation and law enforcement) are becoming an even more attractive target for illicit exploitation by terrorists and hackers [2, 3], increasing the risk of intentional alteration of GNSS signals by means of jamming, meaconing, or spoofing attacks (as previously discussed in Chapter 3). All of these applications and services may need strong countermeasures, possibly based on cryptographically secure signals.

In addition, the need for antispoofing techniques is also motivated by the growing penetration of GNSS technology in commercially sensitive LBSs and liability-critical applications, in which the information about a user's position or velocity is used at the basis for legal decisions or economic transactions (road user charging, pay-as-you-drive insurance, mobile payments, geographic digital rights management, online gambling, and so on). Applications belonging to these types of applications are focused mainly on the reliability (i.e., integrity and authentication) of the estimated position/time, especially in order to avoid possible intentional fraud or mischarging events [4].

Referring to Chapter 3, which presented the characteristics of structured RF interfering signals, this chapter introduces some examples of suitable countermeasures. We focus on methods that are effective against spoofing. Note that the development of spoofing detection and mitigation techniques is an active research topic for the GNSS community and the number of novel approaches and solutions is increasing. However, the aim of this chapter is to provide a general classification of these approaches that is as comprehensive as possible, and to briefly explain some known techniques. Interested readers can refer to the list of references at the end of the chapter in order to obtain additional technical information and implementation details for each approach.

A first macroclassification for antispoofing techniques [5] divides the techniques into *noncryptographic spoofing defenses*, which do not depend on signal encryption or digital signatures, and *cryptographic defenses*, which rely on secret keys that encrypt or digitally sign components of the signals broadcast by GNSS satellites.

First, this chapter describes some common antispoofing methods for stand-alone receivers (Section 8.2), and then it introduces methods that rely on complementary technologies to GNSS (Section 8.3). After that, an overview of possible authentication methods based on cryptographic defenses and suitable to civilian GNSS signals is provided (Section 8.4). The chapter ends with some final remarks (Section 8.5).

8.2 GNSS Receiver Stand-Alone Techniques

Together with the design of encrypted GNSS signals for civilian use, there is a growing interest in stand-alone, receiver-based defenses that process the received signal and determine whether or not it is genuine. Such interest is also driven by the fact that GPS, currently the most used GNSS, does not incorporate authentication means in its civilian signals, due both to institutional priorities and to long procurement and deployment cycles [6].

Generally, most of the antispoofing techniques for stand-alone receivers work at the baseband signal processing level, without attempting to mitigate or remove false signals. A very clear and detailed analysis of spoofing countermeasures was provided by the authors of [3], where some techniques were roughly classified as spoofing detectors (i.e., they discriminate the presence of spoofing signals, without necessarily mitigating the effect of the attack) and spoofing mitigation (i.e., they attempt to neutralize the detected spoofing signals, restoring the correct positioning capabilities of the receiver). Several types of countermeasures have been proposed with different characteristics in terms of complexity, performance, and cost.

In the following subsections, we focus our attention on algorithms based on measurement consistency checks (Section 8.2.1) and methods borrowed from the signal quality monitoring field (Section 8.2.2).

Other antispoofing techniques suitable to stand-alone GNSS receivers and worth mentioning are those based on:

- Spatial processing (i.e., direction-of-arrival comparison with an antenna array, pairwise correlation in a synthetic antenna array, multiantenna beamforming, and null steering);
- Time-of-arrival discrimination (focusing on the PRN code and data bit latency in spoofing signals);
- Distribution analysis of the correlator outputs (monitoring possible fluctuations due to the interaction between the authentic and the spoofing signals);
- Vestigial signal defenses (based on the fact that the spoofer generally does not suppress the authentic signals):
- Receiver autonomous integrity monitoring (RAIM), which is suitable for detecting anomalies in pseudorange measurements.

A comprehensive analysis of all of these techniques is beyond the scope of this chapter. However, for more details on these techniques, interested readers can refer to [3, 6–10].

8.2.1 Consistency Check of Receiver Measurements

Current commercial receivers typically do not include antispoofing measures. The conventional algorithms implemented in mass-market products are not able to discern real and spoofed signals. However, in some cases the internal logic of commercial receivers already acts as a sort of protection against simplistic spoofing attacks. In fact, they tend to exclude satellites that are not supposed to be in view from the position, velocity, and time (PVT) computation.

As explained in Chapter 3, there are different ways to generate counterfeit GNSS signals and carry out a spoofing attack. Some of these do not synchronize the fraudulent devices to the GNSS time scale and, if the receiver if fooled, some of the receiver measurements can have anomalous trends. Effective spoofing detectors can be built on top of simple algorithms that monitor some indicative signal parameters within the receiver.

A possible antispoofing solution can be based on the detection of possible discontinuities (i.e., jumps) on the position estimated by the receiver. Furthermore, with a simplistic spoofing attack, the transition from the real to the false signals induces a discontinuity in the estimated GNSS time that is unexpected in regular GNSS processing. An example of discontinuity in the estimated position and the time scale due to a simplistic spoofing attack was illustrated in Figure 3.2 of Chapter 3.

In addition to the detection of possible discontinuities in the PVT time estimates, spoofing detectors can be based on the monitoring of the received signal power. One of the common output values for a satellite navigation receiver, apart from the PVT triad, is the carrier-to-noise density ratio (usually indicated as C/N_0) measured for each satellite signal, which is the ratio between the satellite signal carrier power and the noise power spectral density at the reception antenna. The receiver continuously monitors the estimated C/N_0 for each satellite and looks for any unusual variation that might be a sign of a spoofing attack [11]. The receiver can discriminate those PRNs whose absolute power is some decibels (e.g., 2 dB) higher than the maximum possible received power. In addition, unusual power variations correlated with the receiver movements can be an additional indicator of a possible spoofing attack; in fact, the relative movement between the target receiver antenna and the spoofer can considerably change the C/N_0 related to the spoofing signals.

The C/N_0 monitoring on one single band, for example, the GPS L1, can be cross-checked with the C/N_0 estimate on a different band, for example, the GPS L2, if the receiver features multifrequency capabilities [12]. This cross-check leverages on the fact that there is a predefined power level difference between GPS signals in different frequency bands.

Considering a GPS L1/L2 receiver, in addition to the C/N_0, it is possible to monitor other receiver measurements, whose trend can be predicted in the absence of interfering signals. For example, an effective detector could be designed by looking at the relative delay of the correlation peak corresponding to a defined PRN on both L1 and L2 [12]. Similar to the case of jamming, the frequency diversity is an intrinsic countermeasure because it forces the spoofer to generate false signals in multiple bands, increasing its complexity and cost.

Note that the current trend of GNSS receivers, including those for mass-market applications, is toward multiconstellation devices [13]. The possibility of having more satellites in view improves positioning performance, but can also be seen as a possible defense against false signals thanks to the redundancy. For example, consistency checks among navigation solutions computed by using subsets of the satellites in view, for example, those belonging to a single GNSS constellation (e.g., GPS stand-alone versus GLONASS stand-alone), would be an option for implementing a spoofing detection. For a multiple-frequency situation, more constellations increase the required complexity of the spoofing devices, which should be able to mimic different GNSSs to be effective.

In addition to these approaches, a number of other solutions have already been proposed in the literature for detecting and mitigating spoofing attacks (see, e.g., [3]). Among them, it is worth mentioning the consistency checks of the code and phase rate measurements (based on the fact that, in the case of authentic signals, the Doppler frequency and the code delay rate are consistent), the GNSS timing information (checking the consistency of the GNSS clocks obtained from different satellites), and the received ephemeris (cross-checking of ephemeris received from different satellites).

8.2.2 Signal Quality Monitoring

Signal quality monitoring (SQM) refers to the implementation of specific algorithms within the receiver to monitor distortions of the received signals. SQM is a simple approach that can be used to raise an alert, if some test metrics do not pass predefined quality levels. SQM was first proposed for integrity monitoring and was used to observe the shape of the signals broadcast by the satellites. Recently it has been suggested as an effective method against spoofing attacks.

SQM algorithms are based on analysis of the correlation function between the incoming signal and a local replica of the pseudorandom code [5]. Traditional receiver architectures are adapted to perform specialized signal processing, thus enabling an indirect monitoring capability in the correlation domain. SQM algorithms are based on a multicorrelator structure that detects possible distortions on the correlation due to irregularities on the received signal. These algorithms employ from three or more correlator pairs per channel, each slaved to the tracking pair. The measurements from each correlator output are used to form detection metrics, which are, in general, simple algebraic

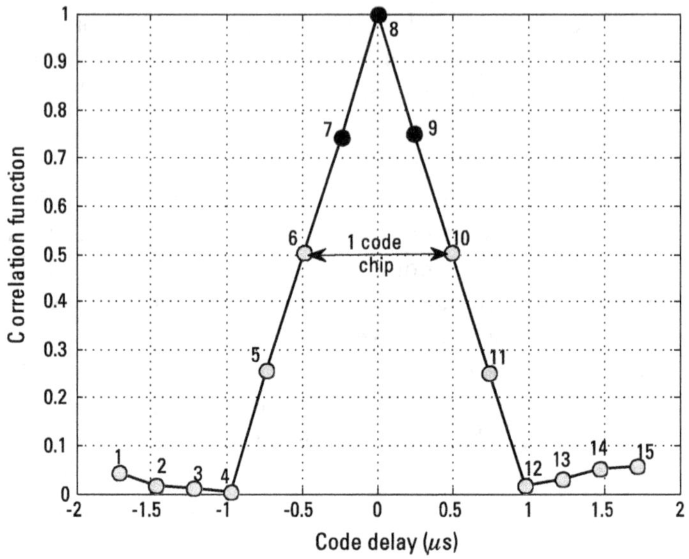

Figure 8.1 Example of multicorrelator structure tracking the GPS C/A code correlation peak.

combinations of the measurements. Figure 8.1 shows the correlation peak associated with a GPS L1 C/A code signal tracked by a multicorrelator structure, with six correlators on the early side (i.e., 1 to 6), six correlators on the late side (i.e., 10 to 15), and three correlators on the correlation peak (i.e., 7 to 9). A correlation pair, for example, correlators 6 and 10 (with a spacing equal to one code chip), is used for tracking purposes and control the other pairs.

Considering Figure 8.1, at the end of each integration period (e.g., 20 ms), a set of 15 correlation values can be used to form a detection metric, which is normally compared against a threshold to determine whether the correlation is distorted or deviates from the nominal shape. As an example, a distorted correlation function due to an intermediate spoofing attack was illustrated earlier (see Figure 3.3 in Chapter 3).

Several metrics for SQM have been proposed in the past, some of them have been reported and compared in [14]. Two common SQM detection tests are the *delta test* and the *ratio test*, which have been considered as spoofing detectors [15] with good performance against intermediate spoofing attacks [16].

The *delta test* identifies asymmetric correlation peaks and is defined as

$$\Delta_m = \frac{I_{E,m} - I_{L,m}}{I_{P,m}} \tag{8.1}$$

where $I_{E,m}$, $I_{L,m}$, and $I_{P,m}$ represent the in-phase early, late, and prompt correlator outputs, respectively; and m is an index to indicate the correlator pair in the multicorrelator structure. The division by $I_{P,m}$ at the denominator serves as normalization of the metric in (8.1), making it independent from the received signal amplitude.

On the other hand, the metric associated with the *ratio test* is defined as

$$R_m = \frac{I_{E,m} + I_{L,m}}{I_{P,m}} \tag{8.2}$$

The ratio test was originally designed to identify flat or abnormally sharp or elevated correlation peaks. An example of the use of the ratio test to detect intermediate spoofing attacks is discussed next, through an experiment performed in a laboratory.

A real GPS signal at the antenna was processed by a modified real-time software receiver that was acting as a spoofer. This modified receiver had the ability to generate false signals starting from the local carriers and spreading codes locked on the signal-in-space (SIS). The counterfeit digital signal samples at the baseband were converted to analog signals, brought to RF, and finally combined with the real signals. After an initial calibration phase, the software receiver adjusted the local carrier frequencies to recover the frequency shift of the oscillator in the front end. In addition, it implemented an algorithm for navigation data bit prediction in order to compensate the delay introduced by the signal processing blocks for the generation of the counterfeit signal.

The composite signal, resulting from the sum of the real and counterfeit ones, emulated an intermediate spoofing attack. The composite signal was sent to a commercial GPS front-end and the raw samples stored in memory were processed off-line with a software receiver implementing the ratio test. To better appreciate the experimental results, Figure 8.2 sketches four temporal phases of the attack in the correlation domain:

- *Initial phase* (Figure 8.2, top subplot). The spoofing signal is active, but still "far" from the real correlation peak, which is tracked by the target receiver.
- *Approaching phase* (Figure 8.2, second subplot). The counterfeit peak approaches the genuine one.
- *Overlapping phase* (Figure 8.2, third subplot). The counterfeit signal distorts the genuine correlation peak and forces a signal lift-off (if the signal power is higher than the real one). The distortion on the

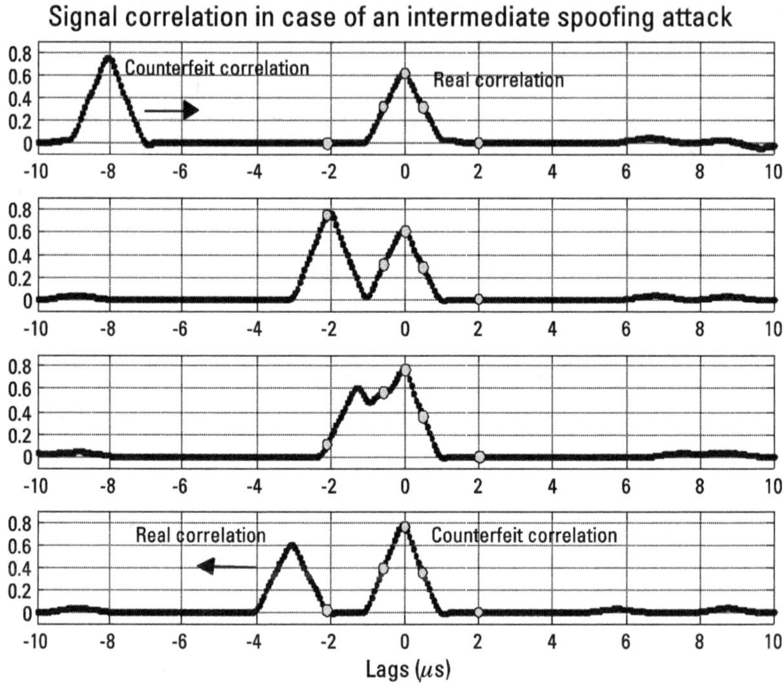

Figure 8.2 Four phases of the intermediate spoofing attack in the correlation domain.

correlation function induced by the spoofed signal can be detected by SQM algorithms.

- *Dropping phase* (Figure 8.2, bottom subplot). The counterfeit signal power is sufficiently high to force the receiver to stay locked on the counterfeit signal, which is delayed in order to introduce a pseudorange error.

Figure 8.3 reports the trend of the ratio test metric versus time during these phases of the attack.

Considering that the ratio test is designed to reveal asymmetries on the correlation function, as expected, its trend became significantly irregular when the real and the counterfeit correlation peaks collided, especially at the beginning and end of the overlapping phase (i.e., from 5s to 10s, and from 20s to 25s in Figure 8.3). The ratio test metric can be compared against a predefined threshold to raise alarms in case of correlation distortions. However, as for any decision process, it is important to derive the probabilities of false alarm and

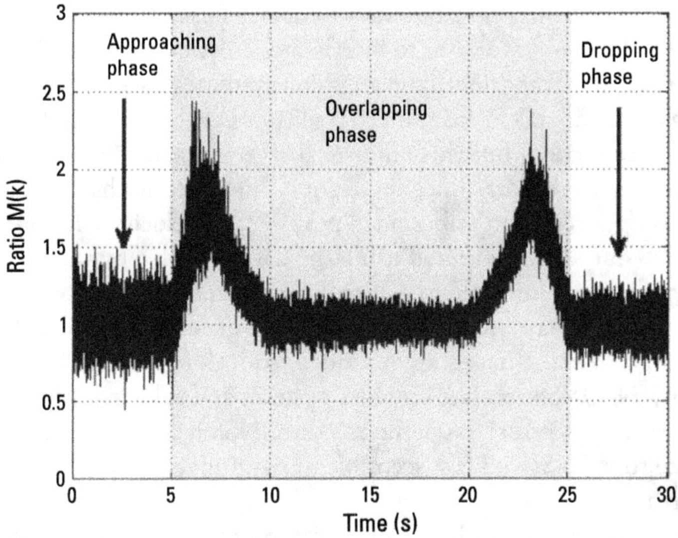

Figure 8.3 Trend of the ratio test metric during the spoofing attack reproduced in a lab.

correct detection. For example, it is possible to adopt the Neyman-Pearson [17] detector, which implements a binary hypothesis test that is able to choose between the null hypothesis and the spoofing-present hypothesis [16].

8.3 Hybrid Positioning Receiver Techniques

This section presents antispoofing techniques that leverage the presence of additional sensors, external to the GNSS receiver. Such sensors provide aiding data that is used to cross-check the GNSS measurements. The diversity introduced by other technologies is not only a barrier against spoofing, but it makes GNSS receivers more robust to any impairment on the RF signal.

8.3.1 Integration with Inertial Systems

Spoofing and jamming devices illegally transmit RF signals on bands allocated to GNSS and, in both the cases, the victim receives a corrupted GNSS signal. Clearly, the integration with external inertial sensors intrinsically forms a barrier against intentional attacks, increasing the robustness of the navigation system. Before detailing possible ways to detect false GNSS signals leveraging on inertial sensors, we want to briefly recall the benefits provided by the integration of GNSS and inertial navigation systems (INSs).

Measurements taken from inertial sensors exhibit a relatively low noise from second to second, but this tends to drift over time due to the inherent error of the inertial sensors. Since this error is integrated in the mechanization process, it is reflected in an unbounded position and velocity solution. Despite this, the INS output is computed using the data provided by the inertial sensors, which makes it immune to external interferences [18].

On the other hand, GNSS measurements are relatively noisy from second to second, but the biases are bounded, so it does not exhibit long-term drift. Thus, the GNSS receiver provides position and velocity estimation with bounded estimation errors. Nonetheless, the GNSS receiver outputs data with a lower rate than INSs and is susceptible to jamming, blockage, interference, and so forth.

The fusion between GNSS and INSs has been implemented in many applications because it provides better performance than does each device's stand-alone operation, which is a consequence of its complementary nature. Basically, GPS and inertial measurements are complementary for two reasons: the characteristics of their errors are different and these are measurements of different quantities. The redundancy that the two systems can provide leads to the following major advantages:

- The INS provides navigation information when GNSS signals are not available.
- GNSS measurements can be used to correct the INS estimates by an integrated navigation filter.

As described in [3] and [15], a GNSS receiver coupled with an inertial measurement unit (IMU) provides protection by effectively cross-checking the receiver's velocity estimates with the integrated IMU's acceleration measurements. The aiding data from the IMU helps the target receiver to discriminate the spoofing threat, for example, through a Bayesian estimator.

As an example, let us assume that a GNSS receiver is integrated with an IMU in a loosely coupled scheme and that it is attacked by an intermediate spoofing device against one of the satellites in view. We also assume that when the attack starts, the receiver is mounted in a static position and already tracks the real signal. If the receiver remains in the same position, the velocity estimates retrieved by the IMU and by the receiver should be null. If, however, the intermediate spoofing attack successfully forces the receiver channel to track the false correlation peak (thus, losing track of the signal transmitted by the satellite), the new values of velocity estimated by the GNSS receiver will no longer be null. Clearly, the IMU continues to provide null values of acceleration and velocity. Such measurement inconsistency provides a valuable piece of information when designing spoofing detectors. The difference (or linear combination) of measurements provided by the IMU and the receiver can be monitored over time, compared against a threshold, and then used to raise alarms as necessary. Of course, the same example can be extended to the dynamic case, assuming a mobile target receiver.

8.3.2 Integration with Communication Systems

In many LBSs, the navigation system installed on board vehicles computes the users' positions with a GNSS and transmits data to a third party via wireless channels. Today, examples in the mass-market sector are quite common thanks to the widespread use of smartphones and consumer devices that have embedded low-cost GNSS receivers. Examples include systems for road user charging, smart parking, pay-as-you-drive insurance, and fleet tracking. Most of these can be considered liability-critical applications, where some users may have a direct interest in fooling the service to obtain economic gain. In these cases the risk of spoofing is not negligible.

In addition to the countermeasures described in the previous sections, hybridization with communication systems is an additional way to increase the robustness of position versus spoofing [3]. In fact, communication systems such as cellular networks or Wi-Fi stations can be used as wireless positioning systems and can be valuable positioning sources for cross-checking GNSS data. In LBSs that involve a communication network, the user device is associated with one radio cell. From the network, it is possible to retrieve the area where the user is supposed to be, through the cell identifier, or estimate the users' positions with ranging measurements based on received signal strength (RSS) or time-of-arrival (TOA) measurements [19]. The trivial criterion is to determine whether the positions from both positioning sources are consistent. If the confidence region of different solutions does not intersect (i.e., the GNSS position is outside the area associated with the cell identifier), there is a high likelihood that a spoofing condition exists.

For a network-based position, the cell identifier alone does not provide sufficient accuracy; however, its size is reduced with the population density, given the large use of mobile phones, and the fixed limit of communication channels in one cell. The same happens with the road network, so that the density of cells is somehow proportionate to the density of roads, and the possibility to differentiate roads by the cells is grossly constant over all places [20].

8.4 Authentication Techniques

A simple definition for GNSS signal authentication is given in [5] as the "certification that a received signal is not counterfeit, that it originates from a GNSS satellite and not a spoofer."

The concept of signal authentication represents a cryptography-based countermeasure to possible spoofing attacks. It requires the presence of a cryptographically secure portion in the received signal (sometimes referred as a *security code* or *digital signature*) and it involves two subtypes of authentication [5]:

- *Code origin authentication:* certification that the security code originates with the GNSS control segment (i.e., source authentication);
- *Code timing authentication:* certification that the security code arrives promptly (i.e., with the correct time of arrival) and intact (i.e., data integrity).

Unfortunately, position authenticity cannot be ensured by the current stand-alone receivers, which solely exploit civil GNSS signals (e.g., those using the GPS L1 C/A signal) [4]. Some solutions have been proposed, but most of them are based on a client–server approach, in which the hidden/unknown attributes of restricted-access GNSS signals (e.g., the military GPS L1 P/Y codes [6], the Galileo Public Regulated Service signals [21], or the Galileo Commercial Service signals [22]) are cross-compared between different locations in order to authenticate the civil (open service) signals.

However, some interesting proposals have been made for implementing authentication solutions for civil applications based solely on GNSS signals [5, 23–29]. For example, modifications of the current civil SIS (or at least of the navigation message content) have been proposed both for the modernized GPS [5] and the Galileo Open Service signals [28]. Thus, it is reasonable to expect that a position authentication mechanism will be provided in the near future within the GNSS signal itself, as an added value of the GNSS system [4]. In this way future receivers will be able to perform a stand-alone assessment of the authenticity of the computed PVT solution, decreasing the need for costly additional sensors or other countermeasures to spoofing attacks (discussed in previous sections).

Note that two important aspects of the GNSS SIS need to be protected from spoofing attacks [28]: the spreading code chips and the navigation message data bits. The core objective of an authentication mechanism within the GNSS SIS is to introduce features that make it difficult for an attacker to generate valid signals. Four techniques based on cryptography are widely referenced in the GNSS literature (see, e.g., [26, 28]) as suitable candidates for authenticating either the navigation message or the spreading codes:

- *Navigation message authentication (NMA):* Denotes the authentication of satellite navigation messages by means of digitally signing the navigation message data and thus keeping the navigation message clear (i.e., unencrypted).
- *Spreading code authentication (SCA):* Embeds short encrypted code segments within the nominal (unencrypted) spreading code sequence.
- *Navigation message encryption (NME):* Refers to the encryption of the whole navigation message, which is then modulated on the spreading codes.
- *Spreading code encryption (SCE):* Denotes the encryption of the whole spreading code sequence transmitted by each satellite.

More details about these four authentication techniques are provided in the following subsections, where implementation aspects are also discussed, including the related complexity of the GNSS system and the receiver, and the robustness to meaconing and spoofing attacks.

8.4.1 Navigation Message Authentication

An important milestone for GNSS authentication was placed by Logan Scott in 2003: antispoofing techniques and authentication mechanisms were discussed in [23], proposing to modify the civil GNSS signal structures in order to incorporate explicit authentication features. After that, the NMA solution was proposed in [24] and since then has been mentioned in many papers in the scientific literature (e.g., [5, 25, 26]). It refers to the authentication of satellite signals by digitally signing the modulated navigation data, typically using asymmetric cryptographic algorithms.

The main concept behind this technique is having a key pair with two distinct keys, one private and one public. The private (signing) key is secret and is only known by the GNSS control segment. The public (validation) key is made public and available to any user of the system (including fraudulent users). The navigation message consists of several blocks of data, containing satellite clock parameters and ephemeris. A digital signature for these blocks is generated by means of a digital signature algorithm (using the private key). The digital signature can be cryptographically generated using past or future blocks of data and it is broadcast as an additional block in the navigation data stream. The user receives the navigation data and the signature via the modulated data bits. After receiving a complete message, the user can authenticate it by using a validation function (knowing the public key).

The robustness of the authentication procedure resides in the cryptographic strength of the public/private key pair; it means that just knowing the public key and past navigation message data, it is computationally infeasible for an attacker to generate a valid digital signature (i.e., to recover the private key).

One important disadvantage of NMA is the existence of a delay in the authentication process. The receiver is only able to authenticate a signal after the reception of a whole portion of the navigation message, including a valid digital signature.

The effect of the authentication delay on the final user application depends strongly on the user and system requirements. In fact, if the GNSS receiver must authenticate the received SIS in real time, the authentication delay is definitely an issue (e.g., a receiver used in a network node for the provision of timing reference signals).

In contrast, for a group of users sending their positions to a remote control center in charge of monitoring their positions, the NMA can be an appropriate solution if the real time is not a severe constraint (e.g., monitoring of a fleet of trucks or fishing vessels).

It is important to note that, since the NMA acts at the navigation message level, it does not ensure the correct time of arrival of the signal: it may fail against meaconing and intermediate or sophisticated spoofing attacks able to decode live signals and replay counterfeit signals in near real time (i.e., SCER attacks, as previously described in Section 3.4.3). To strengthen its robustness (providing both code origin and code timing authentication), a combination of NMA with one or more noncryptographic antispoofing defenses would be recommended [5].

The main advantages and drawbacks of the NMA technique are summarized in Table 8.1.

Table 8.1
Main Characteristics of the Navigation Message Authentication Technique

System complexity	**Low**
Receiver complexity	Low + The authentication process can be performed through software routines and does not require major changes in current GNSS receiver architectures.
Robustness against meaconing/spoofing	+ Robust against simplistic spoofing attacks. – Suffers from authentication delay. – Fails against meaconing and intermediate/sophisticated spoofers able to use live signals and replay counterfeit signals in near real time (i.e., SCER attacks).

8.4.2 Spreading Code Authentication

The SCA method expands the NMA concept, tightly binding an additional security feature in the spreading codes. Different solutions have already been proposed for the implementation of an SCA mechanism. Besides the digital signature on the navigation message, in one proposal [23] additional code segments are inserted into the ranging code in fixed time windows. These are called spread-spectrum security codes (SSSC) and are cryptographically generated as an enlargement of the digital signature of the navigation message in the form of pseudorandom bit sequences [23, 29].

Note that only the GNSS control segment knows what digital signature is going to be sent in the authentication message of each satellite, several minutes before it is actually sent. The digital signature (not yet sent) is used to generate the SSSC sequence, which is interleaved to the ranging codes.

This strategy suitably applies to those signal formats that use data and pilot channels (e.g., the Galileo E1 OS signal). SSSC can be embedded only in one component (e.g., the data channel), while the second (e.g., the pilot channel) can be used by the receiver to track the incoming signal during SSSC intervals, until the authentication process is completed. Alternatively, a cryptographically generated code could be added just as an additional signal component, for example, as an extra pilot channel [28].

On the user side, to complete the authentication procedure, the receiver must be able to store in memory raw signal samples at the output of the A/D converter. In detail, the receiver knows precisely when the SSSC is being received, but has no idea what the actual SSSC sequence is until the authentication message is received. During SSSC intervals, the receiver collects precorrelation samples [23] and stores them in memory. Furthermore, during SSSC intervals, the receiver treats the symbols at the output of the prompt correlator as erasures.

Once the digital signature is received, the receiver generates a local version of the security spreading code, which is then correlated with the raw samples stored in memory. If the SSSC and its correlation features are not detected at the correct power level, the signal is not authenticated.

Spoofers cannot generate counterfeit signals until an entire navigation message is received. The SCA technique is robust to spoofing, unless fraudulent users use directional antennas to raise the signal above the noise floor and observe SSSC chips directly [23, 26, 29]. Although this solution is theoretically possible, it is quite complex and impractical (as previously discussed in Section 3.4.4).

Like the NMA technique, the SCA technique also suffers from authentication delay. In fact, the received signal can be authenticated only after the reception of a complete navigation message, including the digital signature. The time to authenticate could be reduced using modified versions of the SCA approach (e.g., a technique designated as Private SSSC [23, 26]).

The main advantages and drawbacks of the SCA technique are summarized in Table 8.2.

8.4.3 Navigation Message Encryption

In NME a secret key pair is used to encrypt/decrypt the navigation message, typically using symmetric cryptographic algorithms. The NME method provides authentication if the user community is trustworthy (i.e., if the key is kept secret) [26].

This type of authentication requires the encapsulation of the secret key within tamper-resistant hardware, likely increasing the receiver's cost. In fact, an additional module for decrypting the navigation message is required in the receiver architecture. This module could use standard cryptographic interfaces to store the secret key (e.g., a smart card).

The main advantages and drawbacks of the NME technique are summarized in Table 8.3.

8.4.4 Spreading Code Encryption

Signal authentication can be achieved by encrypting the spreading code, typically using symmetric cryptographic algorithms. At the moment, the SCE is the solution adopted for restricted signals, for example, the GPS P(Y) code or the Galileo Public Regulated Service.

Table 8.2
Main Characteristics of the Spreading Code Authentication Technique

System complexity	Low/Medium This technique: – Foresees the use of additional segments of spreading codes (SSSC), interleaved with those used for ranging; + Can be better applied to GNSS signal formats featuring data and pilot channels.
Receiver complexity	Medium/High – To complete the authentication procedure, the receiver must be able to store raw samples at the A/D converter output. – The receiver must be able to process data and pilot channels (risk of loss of lock of the tracking loops during the authentication procedure, if the receiver uses only one channel). – Suffers from authentication delay, like the NMA. – Medium increase in receiver's cost.
Robustness against meaconing/spoofing	+ Robust against simplistic and intermediate spoofing attacks. Successful spoofing attacks might be accomplished only by using high-gain antennas (in order to rise the SSSC above the noise floor), which is logistically complex and quite impractical. – Fails against meaconing.

In the SCE approach, the ranging code used by the satellite is encrypted by modulo 2 addition of a pseudorandom bit sequence. If the chip rate of the encryption stream is identical to that of the ranging code, the modulo 2 addition results in a pseudorandom sequence [26]. The chip rate of the encryption stream can also be slower than the chip rate of the ranging code. In this case, some code sequences result known except for the sign. This limits the possibility of authentication, because such code sequences can be still used, for example, to perform pseudorange code measurements.

To gain full user and signal authentication from SCE, the receiver architecture must include tamper-resistant hardware for storing the secret key and for performing the local code generation. The implementation of such a tamper-resistant module is more complicated than is required for the NME technique. In fact, the whole digital signal processing unit in charge of the code generation has to be secured in the case of SCE.

As summarized in Table 8.4, the SCE technique requires tamper-resistant hardware and secure management of the secret key, which might be concerns

Table 8.3
Main Characteristics of the Navigation Message Encryption Technique

System complexity	Medium – Requires a secret key pair to encrypt/decrypt the navigation message. – Key distribution and management can be an issue.
Receiver complexity	Medium – Requires tamper-resistant hardware to keep secret the key used to decrypt the navigation message. – The tamper-resistant module could be external to the core of the navigation receiver. – Does not need to acquire and track an encrypted spreading code. – Moderate increase in receiver's costs.
Robustness against meaconing/spoofing	+ Robust against simplistic spoofing attacks. – Fails against meaconing and intermediate/sophisticated spoofers able to use live signals and replay counterfeit signals in near real time (i.e., SCER attacks).

Table 8.4
Main Characteristics of the Spreading Code Encryption Technique

System complexity	Medium/High – Requires secret key to encrypt/decrypt the spreading codes. – Key distribution and management can be an issue.
Receiver complexity	High – Requires tamper-resistant hardware to keep secrete the key used for generating the spreading code. – The tamper-resistant module is embedded in the navigation receiver. – Need to acquire and track an encrypted spreading code. – The cost and complexity of the receiver increases significantly.
Robustness against meaconing/spoofing	+ Robust against simplistic, intermediate, and sophisticated spoofing attacks. Successful spoofing attacks might be accomplished only using high-gain antennas to rise the signal above the noise floor, which is quite impractical. – Fails against meaconing.

for commercial applications. However, SCE seems the most effective solution against spoofing attacks. Fraudulent users might be able to break the system only by using high-gain antennas to raise the signal above the noise floor and then by directly observing the spreading code. In principle, this attack is possible, but not practical (as previous discussed in Section 3.4.4).

8.5 Conclusions

A comprehensive overview of antispoofing techniques has been provided in this chapter, including stand-alone approaches based on GNSS signals only, methods relying on complementary technologies to GNSS, and authentication methods based on cryptography.

Note that, if the presented methods are separately considered, none of them can be considered completely effective against all possible menaces (i.e., meaconing and spoofing attacks, as described in Chapter 3). However, a combination of both cryptographic and noncryptographic countermeasures is recommended as the best option in order to strengthen the robustness of a GNSS receiver with a reasonable increase in the receiver's cost and complexity.

References

[1] Dixon, C., et al., "Specification and Testing of GNSS Vulnerabilities," *Proc. European Navigation Conference 2013 (ENC-GNSS 2013)*, Vienna, Austria, April 23–25, 2013.

[2] Wesson, K., and T. Humphreys, "Hacking Drones," *Scientific American*, Vol. 309, 2013, pp. 54–59. doi:10.1038/scientificamerican1113-54.

[3] Jafarnia-Jahromi, A., et al., "GPS Vulnerability to Spoofing Threats and a Review of Antispoofing Techniques," *Int. J. of Navigation and Observation*, Vol. 2012, Article ID 127072, 2012, pp. 1–16. doi:10.1155/2012/127072

[4] Margaria, D., E. Falletti, and T. Acarman, "The Need for GNSS Position Integrity and Authentication in ITS: Conceptual and Practical Limitations in Urban Contexts," *Proc. IEEE 2014 Intelligent Vehicles Symposium*, Dearborn, MI, June 8–11, 2014, pp. 1384–1389. doi:10.1109/IVS.2014.6856485

[5] Wesson, K., M. Rothlisberger, and T. Humphreys, "Practical Cryptographic Civil GPS Signal Authentication," *J. Inst. Navig.*, Vol. 59, No. 3, Fall 2012, pp. 177–193.

[6] Lo, S., et al., "Signal Authentication: A Secure Civil GNSS for Today," *Inside GNSS*, Vol. 4, No. 5, September/October 2009, pp. 30–39.

[7] Psiaki, M. L., S. P. Powell, and B. W. O'Hanlon, "GNSS Spoofing Detection. Correlating Carrier Phase with Rapid Antenna Motion," *GPS World*, Vol. 24, No. 6, June 2013, pp. 53–58.

[8] White, N. A., P. S. Maybeck, and S. L. DeVilbiss, "Detection of Interference/Jamming and Spoofing in a DGPS-Aided Inertial System," *IEEE Trans. on Aerospace and Electronic Systems*, Vol. 34, No. 4, 1998, pp. 1208–1217.

[9] Pini, M., B. Motella, and M. Troglia Gamba, "Detection of Correlation Distortions Through Application of Statistical Methods," *Proc. 26th Int. Technical Meeting of the Satellite Division of the Institute of Navigation (ION GNSS 2013)*, Nashville, TN, September 2013, pp. 3279–3289.

[10] Humphreys, T. E., et al., "Assessing the Spoofing Threat: Development of a Portable GPS Civilian Spoofer," *Proc. 21st Int. Technical Meeting of the Satellite Division of the Institute of Navigation (ION GNSS 2008)*, Savannah, GA, September 2008, pp. 2314–2325.

[11] Jafarnia-Jahromi, A., et al., "GPS Spoofer Countermeasure Effectiveness Based on Signal Strength, Noise Power and C/N_0 Measurements," *Int. J. of Satellite Communications and Networking*, Vol. 30, No. 4, July/August 2012, pp. 181–191.

[12] Wen, H., et al., "Countermeasures for GPS Signal Spoofing," *Proc. 18th Int. Technical Meeting of the Satellite Division of the Institute of Navigation (ION GNSS 2005)*, Long Beach, CA, September 2005, pp. 1285–1290.

[13] Mattos, P. G., "Markets and Multi-Frequency GNSS," *Inside GNSS*, Vol. 8, No. 1, January/February 2013, pp. 34–37.

[14] Phelts, R. E., T. Walter, and P. Enge, "Toward Real-Time SQM for WAAS: Improved Detection Techniques," *Proc. 16th Int. Technical Meeting of the Satellite Division of the Institute of Navigation (ION GPS/GNSS 2003)*, Portland, OR, September 2003, pp. 2739–2749.

[15] Ledvina, B. M., et al., "An In-Line Anti-Spoofing Device for Legacy Civil GPS Receivers," *Proc. 2010 Int. Technical Meeting of the Institute of Navigation (ION ITM 2010)*, San Diego, CA, January 2010, pp. 698–712.

[16] Pini, M., et al., "Signal Quality Monitoring Applied to Spoofing Detection," *Proc. 24th Int. Technical Meeting of the Satellite Division of the Institute of Navigation (ION GNSS 2011)*, Portland, OR, September 2011, pp. 1888–1896.

[17] Kay, S. M., *Fundamentals of Statistical Signal Processing: Detection Theory*, Vol. II, Upper Saddle River, NJ: Prentice-Hall, 1998.

[18] Garcia Quinchia, A., Performance Enhancement MEMS Based INS/GPS Integrated System Implemented on a FPGA for Terrestrial Applications, Ph.D. thesis, Universitat Autònoma de Barcelona, 2014.

[19] Dardari, D., E. Falletti, and M. Luise, *Satellite and Terrestrial Radio Positioning Techniques: A Signal Processing Perspective*, Boston: Elsevier Academic Press, 2012.

[20] Bardout, Y., "Authentication of GNSS Position: An Assessment of Spoofing Detection Methods," *Proc. 24th Int. Technical Meeting of the Satellite Division of the Institute of Navigation (ION GNSS 2011)*, Portland, OR, September 2011, pp. 436–446.

[21] Rügamer, A., et al., "Privacy Protected Localization and Authentication of Georeferenced Measurements Using Galileo PRS," *Proc. IEEE/ION Position Location and Navigation Symposium (PLANS 2014)*, Monterey, CA, May 2014, pp. 478–486.

[22] Pozzobon, O., et al., "Open GNSS Signal Authentication Based on the Galileo Commercial Service (CS)," *Proc. 26th Int. Technical Meeting of the Satellite Division of the Institute of Navigation (ION GNSS 2013)*, Nashville, TN, September 2013, pp. 2759–2768.

[23] Scott, L., "Anti-Spoofing & Authenticated Signal Architectures for Civil Navigation Systems," *Proc. 16th Int. Technical Meeting of the Satellite Division of the Institute of Navigation (ION GPS/GNSS 2003)*, Portland, OR, September 2003, pp. 1543–1552.

[24] Wullems, C., O. Pozzobon, and K. Kubik, "Signal Authentication and Integrity Schemes for Next Generation Global Navigation Satellite Systems," *Proc. European Navigation Conference (ENC-GNSS 2005)*, Munich, Germany, July 2005.

[25] Hein, G. W., et al., "Authenticating GNSS: Proofs Against Spoofs, Part 1," *Inside GNSS*, July/August 2007, pp. 58–63.

[26] Hein, G. W., et al., "Authenticating GNSS: Proofs Against Spoofs, Part 2," *Inside GNSS*, September/October 2007, pp. 71–78.

[27] Lo, S. C., and P. K. Enge, "Authenticating Aviation Augmentation System Broadcasts," *Proc. IEEE/ION Position Location and Navigation Symposium (PLANS 2010)*, Indian Wells, CA, May 4–6, 2010, pp. 708–717. doi:10.1109/PLANS.2010.5507223

[28] De Castro, H. V., G. van der Maarel, and E. Safipour, "The Possibility and Added-Value of Authentication in Future Galileo Open Signal," *Proc. 23rd Int. Technical Meeting of the Satellite Division of the Institute of Navigation (ION GNSS 2010)*, Portland, OR, September 2010, pp. 1112–1123.

[29] Kuhn, M. G., "An Asymmetric Security Mechanism for Navigation Signals," *Information Hiding, Lecture Notes in Computer Science*, Vol. 3200, Berlin: Springer, 2005, pp. 239–252.

About the Authors

Fabio Dovis holds a Ph.D. in electronics and communications engineering from Politecnico di Torino, Italy, and he is currently an assistant professor in the Department of Electronics and Telecommunications of Politecnico di Torino. He strongly contributed to the creation within the department of the research group Navigation Signal Analysis and Simulation (NavSAS), which specializes in telecommunication skills in the satellite navigation field. During the past 15 years his research has addressed several advanced signal processing topics related to satellite navigation, including digital receivers architecture design, multipath mitigation, interference and spoofing detection, and cooperative positioning. He has acted as an external technical advisor for the European Global Navigation Satellite Systems (GNSS) Supervisory Authority and the European Commission for monitoring ongoing projects and participation with working groups. In June 2011 he was appointed as a member of the Mission Evolution Advisory Group of the European Commission. This group of 23 experts has the task of proposing and evaluating evolutions of the mission objectives for the modernization of the European satellite navigation programs Galileo and EGNOS. He is a member of the steering committee of the conference series *International Conference on Localisation and GNSS*, and he has been a member of several other scientific committees for other international conferences. He is also involved in educational activities in the field of satellite navigation. He is a member of the scientific committee of the Specializing Master on Navigation and related Applications of Politecnico di Torino and coordinated activities at the European level for training and education for Ph.D.s and professionals in the GNSS field.

Emanuela Falletti is the head of the GNSS Core Algorithms Unit in the Navigation Technologies research area of the Istituto Superiore Mario Boella (Torino, Italy). She obtained her Ph.D. in electronics and communications engineering from Politecnico di Torino in 2004.

Her research interests have covered array signal processing, multiantenna systems, and wireless propagation channel modeling; currently they focus on digital signal processing techniques and algorithms for advanced GNSS software receivers, in particular for interference detection and mitigation, multipath mitigation, and signal simulation.

She has been involved in educational activities at Politecnico di Torino, including courses on statistical signal analysis, design and simulation of wireless communications systems, and postgraduate courses on GNSS digital receivers.

She is the coauthor of several scientific papers for journals and international conferences, and acts as a peer reviewer for various scientific publications.

Davide Margaria received a master's degree in telecommunication engineering at the Politecnico di Torino (Italy) in 2007. He is a member of the NavSAS group and he currently works at the Istituto Superiore Mario Boella (ISMB) as a researcher in the navigation technologies area (GNSS Core Algorithms Unit). His research activities are focused on the design and assessment of innovative signal processing strategies tailored to Galileo and modernized GPS signals. He is also in charge of tutoring activities related to the GNSS Introduction course (Master on Navigation and Related Applications of Politecnico di Torino and ISMB).

Beatrice Motella is currently a researcher at the Istituto Superiore Mario Boella, working in the GNSS Core Algorithms Unit within the navigation technologies area.

Her research activities cover different aspects of the GNSS technologies, focusing on the design of algorithms, signal processing techniques, and improvements in GNSS receiver performance. Within this sphere, her main research topic is radio-frequency interference (RFI) monitoring for GNSS receivers and local components specifically designed for safety-of-life (SoL) applications.

She received a master's degree in communications engineering in January 2003 at Politecnico di Torino. In December 2007, she completed the Ph.D. program in electronics and communications engineering at the same university, working in the NavSAS group. During the Ph.D. program she spent almost one year as a visiting researcher at the Satellite Navigation & Positioning Laboratory (SNAPlab) at the University of New South Wales in Sydney, Australia.

Luciano Musumeci is a postdoctoral researcher in the Department of Electronics and Telecommunications of Politecnico di Torino and since January 2011 he has been part of the NavSaS group. His research interests cover assessment of the radio-frequency interference environment within the framework of the GNSS, with a focus on the aviation safety-of-life application.

Research activity performed during his Ph.D. studies focused on the investigation, assessment, and software implementation of innovative signal processing techniques suitable for detection and mitigation of different types of interfering signals, such as pulsed interference and narrowband interference. His research activity also covers the field of high-sensitivity GNSS receiver design in denied environments, where he investigates different assistance data schemes, including Doppler, Doppler rate, and data bit transition information. In the framework of high-sensitivity GNSS receivers he has also started to investigate principles of peer-to-peer and cooperative positioning implemented by sharing assistance and observable data among users.

Marco Pini heads the activities in the navigation technologies research area at the Istituto Superiore Mario Boella (Torino, Italy). He received a Ph.D. in electronics and communication engineering in December 2006 from Politecnico di Torino University. As a result of the experience gained in the design of GNSS algorithms and applications, he has been responsible for several research projects and has been a coordinator of European programs. He acted as a technical reviewer for the European GNSS Agency and is a member of the ETSI Specialist Task Force 474 on the Interoperability and Integration of GNSS with Telecommunications Systems for the Provision of Location-Based Services.

Index

S

The GNSS Technology and Applications Series

Elliott Kaplan and Christopher Hegarty, Series Editors

A-GPS: Assisted GPS, GNSS, and SBAS, Frank van Diggelen

Applied Satellite Navigation Using GPS, GALILEO, and Augmentation Systems, Ramjee Prasad and Marina Ruggieri

Digital Terrain Modeling: Acquisition, Manipulation, and Applications, Naser El-Sheimy, Caterina Valeo, and Ayman Habib

Geographical Information Systems Demystified, Stephen R. Galati

GNSS Applications and Methods, Scott Gleason and Demoz Gebre-Egziabher

GNSS Interference Threats and Countermeasures, Fabio Dovis, editor

GNSS Markets and Applications, Len Jacobson

GNSS Receivers for Weak Signals, Nesreen I. Ziedan

GNSS for Vehicle Control, David M. Bevly and Stewart Cobb

GPS/GNSS Antennas, B. Rama Rao, W. Kunysz, R. Fante, and K. McDonald

Introduction to GPS: The Global Positioning System, Second Edition, Ahmed El-Rabbany

MEMS-Based Integrated Navigation, Priyanka Aggarwal, Zainab Syed, Aboelmagd Noureldin, and Naser El-Sheimy

Navigation Signal Processing for GNSS Software Receivers, Thomas Pany

Principles of GNSS, Inertial, and Multisensor Integrated Navigation Systems, Second Edition, Paul D. Groves

Server-Side GPS and Assisted-GPS in Java™, Neil Harper

Spread Spectrum Systems for GNSS and Wireless Communications, Jack K. Holmes

Understanding GPS: Principles and Applications, Second Edition, Elliott Kaplan and Christopher Hegarty, editors

Ubiquitous Positioning, Robin Mannings

Wireless Positioning Technologies and Applications, Alan Bensky

For further information on these and other Artech House titles, including previously considered out-of-print books now available through our In-Print- Forever® (IPF®) program, contact:

Artech House Publishers
685 Canton Street
Norwood, MA 02062
Phone: 781-769-9750
Fax: 781-769-6334
e-mail: artech@artechhouse.com

Artech House Books
16 Sussex Street
London SW1V 4RW UK
Phone: +44 (0)20 7596 8750
Fax: +44 (0)20 7630 0166
e-mail:
artech-uk@artechhouse.com

Find us on the World Wide Web at: www.artechhouse.com